The Black Pit ...and Beyond

by

J. GORDON MUMFORD

Published by

GENERAL STORE
PUBLISHING HOUSE

Box 28, 1694 Burnstown Road,
Burnstown, Ontario, Canada K0J 1G0
Telephone (613) 432-7697 or 1-800-465-6072

ISBN 1-894263-19-7
Printed and bound in Canada

Cover photograph: Boatload of seamen from a torpedoed merchant ship. (National Archives of Canada, C-054474)
Background photograph: View from aircraft of a convoy of merchant ships assembling in the North Atlantic. (National Archives of Canada, PA-115005)

Layout and Design by Derek McEwen
Printing by Custom Printers of Renfrew Ltd.
General Store Publishing House
Burnstown, Ontario, Canada

Canadian Cataloguing in Publication Data

Mumford, J. Gordon, 1925-
The black pit . . . and beyond

ISBN 1-894263-19-7

1. Mumford, J. Gordon, 1925-
2. World War, 1939-1945–Personal narratives, Canadian.
3. World War, 1939-1945–Atlantic Ocean.
4. Merchant mariners–Canada–Biography.
5. Radio operators–Canada–Biography. I. Title.

D811.M85 2000 940.54'8171 C00-900204-9

DEDICATION

To my wife and partner, Barbara, with love and appreciation. This book would not have been written without her moral support and encouragement. Her assistance and expertise —from computer knowledge, editing, drafting, and researching, to endless hours of printing—have been invaluable.

FOREWORD

Gordon Mumford's account of his experience as a merchant seaman in *The Black Pit . . . and Beyond* is a useful contribution to WW II merchant service history.

The merchant seamen faced harrowing encounters with U-boats, surface raiders, aircraft and mines. If they survived a sinking, the deadly, inhospitable, freezing and unpredictable North Atlantic awaited them. These men showed uncommon valour in the war effort, and it was done without pomp or fanfare. They delivered the goods against overwhelming odds.

For six long years of war they paid the price. The British lost one in four, the Canadians and Australians one in eight. They were targeted for death by Grand Admiral Dönitz, who instructed his men to take no prisoners. He knew the merchant seamen were fundamental to the war effort. They carried the essential war materiel and foodstuffs for Britain and her allies, particularly Russia, through the murderous Murmansk run. They carried troops through a war zone to a war zone, constantly moving through dangerous waters. They raised the siege of Malta.

The enemy's determination to cut the supply lines imperilled the prospect of victory and resulted in massive losses. British Prime Minister Winston Churchill wrote, "The only thing that ever really frightened me during the war was the U-Boat peril . . . I was even more anxious about this battle [to maintain the Merchant Navy] than I had been about the glorious air fight called the Battle of Britain."

Admiral Leonard Murray concluded, "The Battle of the Atlantic was not won by any Navy or Air Force, it was won by the courage, fortitude and determination of the British and Allied Merchant Navy."

Foster Griezic
Adjunct Professor, Canadian History, Carleton University
Advisor/Consultant, Merchant Navy Coalition for Equality

ACKNOWLEDGEMENTS

Thanks to my fellow members of the (Vancouver, BC) West End Writers Club for their invaluable critiques during the preparation of this book.

I would also like to pay tribute to James Crook, second mate of the *Scottish Heather* in December 1942. I owe my life to him, as do my shipmates.

TABLE OF CONTENTS

PART ONE

The Western Approaches

THE ROCKET, with its long thin trail, soars into the night sky and bursts in a flash of light, releasing a flare. Dangling on its parachute, the flare floats gently down on the outer edge of the convoy, illuminating the sea below. "Flare, Mr. Mate, flare!" I shout to the mate in the wheelhouse. My voice sounds strange in my ears as excitement, tinged with fear and apprehension, surges through me. I stare in awe as, one after another, flares begin to mushroom in the sky on our starboard beam.

The mate, alerted by my shout, rushes out of the wheelhouse, takes one look, and yells, "Shit!" Turning quickly, he hits the alarm button on the bulkhead before sprinting to the lanyard outside the wheelhouse. He yanks it hard, releasing the firing pin, and, to the accompaniment of the shriek of the ship's siren, sends the snowflake canister soaring high in the air. It bursts with a brilliant white light illuminating the ocean and ships beneath it, as it, too, floats down on a tiny parachute. The whole convoy is lit up, transforming the scene from darkness to one of daylight clarity. The flares (commonly called snowflakes) cast an eerie light over the turbulent sea creating dark shadows cast by the masts and overhanging superstructures of the ships.

Snatches of distant voices reach my ears. Men, roused from their cabins and mess decks by the strident notes of the ship's klaxon, emerge from the ship's interior. "Come on, Nobby," a seaman shouts, "get off yer bloody ass. Something's up."

The probing U-boat, suddenly exposed on the surface, takes evasive action and disappears into the night. An unnerving silence falls over the convoy, broken only by the rhythmic thud of the ship's engines and the blast of depth charges. The commodore ship, *Empire Shackleton*, breaks radio silence to issue the urgent warning: *Stop firing snowflakes.*

The last snowflake floats down and is extinguished in the sea. Darkness falls once more over the convoy. I no longer think of leaving the bridge for the warmth of my cabin. My senses are fully alert, and I doubt if I can sleep; there are only three hours left before I go on watch. Mine is the graveyard watch, midnight to four.

The sounds and sights of the action still tingle in my bones as, mesmerized, I stare into the night, my nerves taut and expectant. Only the wind howls back at me out of the darkness. The merchant ships are spread in orderly columns over five miles across and one and a half miles deep.

At seventeen, I'm the third radio officer (sparks) on the tanker *Scottish Heather*, bound for New York en route to Curacao to collect a cargo of high octane. The tanker is in ballast, with eight hundred tons of fuel oil for refuelling the escort at sea. This new technique will extend the cruising range of our naval escort.

An image of the second mate forms in my mind, and I smile because I like him. He's an older man, in his late thirties or early forties, tall and wiry with a kind face which lights up when he talks. His name is Crook, Mr. James Crook. He's easy to talk to and doesn't mind talking to teenagers. He's a north country man and his accent reminds me of my mother; she's from the north, too.

I remember him discussing the refuelling operation. "Aye, lad," he said, "if the refuelling is successful, the escort won't have to go to the Azores or Iceland for fuel, but can stay up in the North Atlantic to help guard the convoy." He pauses, shakes his head, then says with a sigh, "Unfortunately, that means this bloody tanker will be a prime target for the U-boats. If they get us, the convoy's defensive screen will be weakened."

It was a night like this, only wilder, when he told me about the *black pit*; I remember distinctly because it was Christmas Eve and the storm was subsiding. The convoy had just re-formed after being scattered in a vicious Atlantic storm for over twenty-four hours. His face was a grey indistinct shape against the background of the night, but it is his voice that I remember.

"We're going into the black pit in a day or two," he said, then paused, sunk in thought, before continuing. His voice reflected his concern. "We've lost a lot of ships there. The U-boats can operate freely on the surface without fear of attack from the air. This may well be a rough patch we're getting into."

He looked at me, his voice grim, yet soft. "Listen, son," he said, "you stick close to me if there's trouble. Our lifeboat's on the aft deck, number three boat on the starboard side. Now, don't forget."

I nod my head; there's a lump in my throat because I feel that he cares. In a way, he reminds me of my father. *Dad, why did you die?* my mind cries out. I bite my lip to force the tears back: men don't cry. My thoughts are chaotic, coming in short bursts. *He's dead.* It's still fresh in my mind and I see his body lying on the couch in Grandma's dining room. I tried to touch him to wake him, but he wouldn't wake. I couldn't understand. *He died of a wasp sting! A silly, stupid little wasp stung him on the lip as he bit into a jam sandwich.* He died quickly, just as long as it took to pull the wasp off his lip, go to the old Dover stove, and throw it in the fire. Dad returned to his chair and died, right there in the room, right in front of Grandma. That's four months ago, about a month before I went to sea. *Why did he have to die just when I needed him?*

My thoughts return to the second mate. *Perhaps he has sons of his own.* A feeling of loneliness comes over me, and I clutch at straws. I need companionship and understanding, someone to talk to.

A sudden rain squall hits the bridge and brings my wandering thoughts back to reality. The bitter cold of the North Atlantic bites into my face and I pull the hood of the duffel coat closer around my head. It is midwinter, 2100 hours on the night of December 26, 1942, and the convoy is now in the black pit area. When I came off watch this afternoon our position was roughly latitude 49 degrees north and longitude 20 degrees west.

The second mate and the captain are with the mate on the bridge. I can hear the low murmur of their voices, barely audible over the hiss and crash of the bow waves as we plough through the seas. The voices become distant and fade away as the men enter the wheelhouse and go into the chartroom.

The bridge is my refuge from the seasickness which has plagued me since I joined the Merchant Navy. Here, with the wind and cold rain in my face, I find a measure of relief from the wrenching spasms of nausea brought on by swaying cabins and the clogging smell of steam radiators. As I peer over the weatherboard, the hood of my duffel coat falls back exposing my head to the elements, but I do not mind.

The voice of the helmsman disturbs my thoughts. "Aye, Sparks," he says, "'tis the seasickness that you'll be having, is it now?" The lilting Welsh accent cuts through the howl of the wind in the ship's halyards, and I laugh.

"Yeah," I reply. "It's better out here in the fresh air, than in the cabin with the smell of those bloody heaters."

I hear him chuckle and then call goodnight. "Take care," he says. The helmsman has been relieved at the wheel by a shipmate and I watch his shadowy form pass down the bridge gangway to the boat deck below.

This is my second voyage; the first was on the *Soborg*, a dirty Danish collier. I'd sailed to Iceland where we bunkered the ships of a Murmansk convoy in Hvalfjord, a lonely barren fjord to the west of the capital, Reykjavik. The storms north of Scotland were frightening. The waves seemed to tower over the small collier, threatening to engulf us. We seemed to bear a charmed life because at the last moment we always righted ourselves to sail up the immense watery slopes, or down into the gullies between. It was a nerve-wracking introduction to the sea for a raw country boy who'd hardly seen the sea, let alone sailed on it. I still haven't got over seasickness. It's not so severe after a few days, but there are times when I just don't care whether I live or die.

Now, as I stand on the bridge with the wind chilling my face, the sickness has gone and I smile as I remember a similar bridge on the collier, my first day at sea. My face a ghostly white, I had dashed to the wing of the bridge, heedless of whether it was port or starboard, windward or leeward. Unfortunately, I chose the wrong side and, uncaring, drooped over the bridge parapet and let go. As the sickness subsided, I sighed with relief, then winced when I saw the mess. "Oh, no!" I groaned. I'd been sick all over the starboard light. The light was no longer green. With a grimace, I moved away hastily, hoping no one had seen me.

The wheelhouse door opened against the blustering wind and slammed shut as the captain, binoculars in hand, came out onto the bridge. He moved over to the wing and stood still for a moment, his body leaning against the bridge weatherboard as he read the signal flags of the commodore ship pitching and tossing in heavy seas. Then, for some reason he glanced down and saw the light. His startled face changed expression. Flushed with anger, he roared, "Who's been sick over the bloody starboard light?"

Propped up against the bridge bulwark, I heard his voice. It pierced my sickened senses as I stared blearily at the heaving and tossing North Sea. I shuddered and cringed visibly. "Sparks," he yelled, "did you do that to my goddamned light?"

He could only mean me. Sparks is the name coined aboard ship for radio men, like the term Chippy for a carpenter. I nodded weakly and muttered, "Sorry, sir, just couldn't help myself. I . . . I feel awful."

He glowered at my cowering figure, with not an ounce of pity for the wretched youth in front of him. "Get a bucket of water and a mop, then clean that bloody light off," he grunted, "and next time if you've got to be sick, do it on the leeward side of the vessel."

Groaning to myself, I hurried to fetch a mop and bucket. Then, dangling precariously over the side of the bridge, I struggled to clean the offending light. It was a bad start to the budding sea career of a young radio officer. In the throes of seasickness, I longed for the solid feel of earth beneath my feet.

As the evening progresses, more and more sightings of U-boats are reported by the escort. The bridge officer grunts, his face set. "There's a hell of a lot of signalling between ships and escort out there, Sparks," he says, joining me on the wing of the bridge. While we lean against the parapet, our eyes pick out the distant pinpoints of light flashing between the escorting naval vessels on the outer fringes of the convoy. "Looks like something's up."

"Yeah," I reply, "but I can't read the signals clearly unless they're pointing in this direction. They look like U-boat contact reports."

He nods his head. "I've seen it before," he says bitterly. "It's probing action. They're trying to find our weak spots. God help us when they do." He pauses a moment, then adds, "Those U-boats are faster on the surface than the corvettes. They can outrun the escorts any time."

"D'you think they'll attack tonight?" I ask.

"Well," he says, "judging by all the radio reports there must be at least three wolf packs around the convoy. It's got to come some time." There's silence for a moment as he pauses, gathering his thoughts. "They've been shadowing us for three days now, calling up other U-boats. They won't wait much longer. The captain's getting some rest while he can; it's going to be a busy night." He looks at me and adds, "Tomorrow around sixteen hundred hours, we've got a corvette, the *Chilliwack*, to refuel and that's going to be a bloody hassle in these seas."

He rubs his eyes with the back of his hand and peers into the darkness. His voice sounds tired as he continues. "You should get some rest, too. You've got the midnight watch, haven't you? Go to bed for a couple of hours; it'll do you good, son. The watchman will call you."

Reluctantly, I agree. *Who knows what the night will bring?* I make my way to my cabin, leaving the bridge to the mate. The heavy weather door swings open as I push the metal lever down and step inside. A thick blackout curtain, interleaved with another, hangs on a sliding rail, and parts as I enter the brightly lit passageway. My cabin is close to the radio room on the boat deck and shares the same short passageway. A flight of steps at the far end of the passage leads up to the wheelhouse. Compared to the cramped quarters on the *Soborg*, my previous ship, my cabin is luxurious. Though not large, it has its own cupboard, settee, washstand, mirror and bunk. A plain green carpet with a thick pile covers the deck and beige curtains hide the porthole. There is a small writing desk in one corner.

The ship rolls violently in a heavier than usual sea, and I grasp the bunk for support, bracing my legs against the roll. I shake my head. The roll and pitch of the ship remind me of the *Chilliwack*. I remember the problems with the destroyer, *St. Laurent. The mate's right; it's not going to be easy, not in this weather.* We'd tried refuelling the Canadian destroyer yesterday, Christmas Day, but it was unsuccessful. Heavy seas exploding on the destroyer's bow sheared the for'ard hawser, forcing the destroyer to swing wildly off.

A shiver runs through my body as I remember the scream of that steel stranded wire slashing through the air. Even I, on the safety of the bridge, instinctively ducked at the sight and sound of the severed hawser. "Fucking hell," a crewman's voice had shouted in panic, "hit the deck!" and the deck crew dived for cover. Fortunately, the hawser did not hit anyone, because it could have easily cut a man in two. It fell partly on the deck and partly in the sea.

With the operation aborted, we had followed in the destroyer's wake at our maximum speed of twelve knots to rejoin the slow-moving convoy, hull down on the horizon. Earlier today, the tanker again tried to refuel the *St. Laurent*, this time with more success. One hundred tons of fuel oil were pumped into her before the hawser broke a second time.

It's just over two weeks now since I joined the *Scottish Heather* in North Shields, downriver from Newcastle on December 11. The next day, in dirty weather, we sailed in a convoy of twelve ships round the coast of Scotland. Two other convoys from the Clyde and southern coastal ports joined us off Malin Head, Ireland, in the Atlantic convoy assembly area. From these three convoys on the afternoon of December 19, 1942, convoy *ONS 154* was born. Forty-five merchant ships with their Canadian escort of five corvettes, *Chilliwack*, *Shediac*, *Napanee*, *Battleford* and the *Kenogami*, from the Canadian convoy escort *Group C1* moved off in formation westward past Bloody Foreland (a barren lonely headland with a derelict lighthouse on the northern coast of Donegal, Ireland) into the Atlantic Ocean. The Canadian destroyer *St. Laurent* joined us a day later.

On the night of December 22, a gale hit the convoy with wind force nine and mountainous seas, scattering the convoy as each ship struggled for survival. We made little or no headway for twenty-four hours. The storm continued all through the daylight hours of December 23 and into the night. It eased during the early hours of the morning. With the barometer rising, the wind fell to force five. The scattered convoy regrouped, and, in heavy seas, ploughed on westward under overcast skies and rain squalls. It was Christmas Eve, my first Christmas away from home. So much has happened since then.

I hesitate as I take off my outer clothing, trying to decide whether I should get into my pyjamas, but decide otherwise and lie on the bunk, half dressed. The mere act of lying down soothes my stomach and the need to vomit subsides. I doze off into a deep sleep.

A hand shaking my shoulder wakes me. I groan, not fully awake. It seems only a few minutes, yet two hours have passed since I lay down. The bridge officer has sent one of his crew to waken the next watch.

"Wake up, Sparks. It's ten minutes before your watch. Christ, you sleep bloody soundly; I thought you were dead. Here, I've brought you some coffee."

Rubbing the sleep from my eyes, I sit on the edge of my bunk and sip from the mug of steaming coffee. My mind wanders; I long for the feeling of security of our creaky old farmhouse and the comfort of familiar things. *My younger brothers will be asleep now. Mother packs them off to bed about nine in the evening. As for Dad? Well, he's dead; I won't see him again.* A feeling of helplessness comes over me; I put away these thoughts and wash quickly. It's time to go on watch.

As I enter the radio cabin, Dick stretches his arms above his head and yawns. "God, I'm tired, Gord." He signs off the log and gets up from the operating chair. "Aye, lad," he says, "it's getting rough out there by the sound of it. T' bloody U-boats are all around us. Here, I've had enough. It's all yours. The graveyard watch! What a goddamned name."

Short and dark-haired, Dick's in his mid-thirties and in charge of the radio section. He's from the Newcastle area, a Geordie, while John, the second sparks, and I are both from the south. We're on a first name basis, except when others are present. Then we simply call him Chief. His parting words linger in my mind as he leaves. *Yeah, the watch is aptly named*, I think to myself; *if anything happens it will be in the dead of night.*

Settling down in the operating chair, I reach for the headphones. My eyes automatically check for the little red can, my constant companion in the radio cabin. It's close by if I need it. Seasick or not, the job has to be done. An uneasiness takes hold of me. I feel edgy and confined in the cabin, not knowing when the U-boats will attack.

The medium wave receiver is monitoring the 500 Khz distress band, while my fingers rapidly spin the HF (high frequency) receiver dial to Rugby Radio. It is time for the regular radio broadcast schedules for merchant shipping. I concentrate as I listen to the long list of ship and convoy call signs. The static is heavy tonight, making it difficult to read the Morse. There're several messages for the convoy; it looks like I'm going to be busy receiving and decoding them before handing them to the officer of the watch.

The captain is on the bridge with the mate; an attack is inevitable. The U-boats are launching probing attacks, and the distant thuds of bursting depth charges unsettle me. My body tenses as the vibrations shake the ship and my imagination plays tricks on my mind. I feel my skin tingling in expectation of the shattering explosion of a torpedo hitting the ship. It doesn't happen, and the thuds die away.

My contact with the outer world is through the HF radio. Sealed into the radio cabin, my ears glued to the radio for the four hours of the watch, I get another dimension on the war as the U-boats continue probing for weak spots. I force my mind away from the noise and vibrations. The shrill chatter of Morse runs through my head as I automatically copy out the messages for the convoy and decode them. Setting them aside for the mate, I ring the bridge on the intercom network. A few moments later the door opens and he pokes his head round the door. "You've got something for me, Sparks?"

I grimace and nod, "Not good news, I'm afraid. I've decoded a couple of messages. It looks like the number of U-boats out there has increased."

The mate's face is serious as he takes the messages. "Yeah, I was afraid of that. Things are hotting up now." He looks up from the papers in his hand. "It must be nerve-wracking in here, son," he says. "At least we can see, on the bridge." Pausing as he turns toward the cabin door, he adds, "I'll come in occasionally, and if you need help decoding, then I'll get someone to help you. The Old Man's with me on the bridge. At the moment there's only probing action by the U-boats."

Attack: The Black Pit

THE ATTACK against the convoy comes around 0130 hours, halfway through my watch. Swiftly, and without warning, a violent explosion from a nearby vessel shakes our ship. A distress call follows almost immediately on the 600 metre band. My fingers tremble as I log the call at 0140 hours on December 27. The radio officer on the stricken ship repeatedly sends the call, *SSS de 121 torpedoed*, in Morse code before he, too, abandons ship. Only the ship's convoy position number is used, not the ship's own call sign. The figures "121" mean twelfth column and first ship (leader) of the column. It is the *Empire Union*, on our starboard flank.

The sinking brings the war closer, if that is possible, but there is no time to dwell on it. "Jesus Christ!" Startled, my pen jumps on the message pad as, just minutes later, the blast from another explosion hits the tanker, making her heel over violently. It sounds even heavier and nearer than that of the *Empire Union*. "Bloody hell," I mutter, and nervously bite my lip. "That's too bloody close."

The intensity of the action dies down, and only the occasional distant thuds of depth charges break through the steady beat of the ship's engines. The cabin door opens and the mate steps in. I glance up. He's a large man in his mid-forties, and his face shows the weariness and tension of the attack. His duffel coat is wet and the hood has fallen back revealing the slicked-back hair and receding temples. "Everything okay, Sparks?"

I nod. "Yeah. I've logged the distress call from the *Empire Union*, but a few minutes later there was another hell of an explosion. What happened? There was no distress call."

His hand smooths back his greying hair, "It was the *Melrose Abbey*, the leading ship in column ten. Her back was broken and her bridge on fire." He pauses, his voice thick, "I don't think the sparks had a chance to send the distress. She went down in a matter of minutes."

Reassured that all is well with me, the mate hurries back to the bridge. I settle in my chair turning over his words in my mind as I monitor the frequencies. *Two ships down, and this is just the start.*

At four o'clock, John takes over the watch from me. He looks drawn and tired. He had been aroused from a deep sleep when the attack started. "Christ, Gord," he mutters, "What a bloody night. The klaxon woke me. I managed to doze off for a bit on the settee about two-thirty, but I feel lousy."

I stand up and look down at the log book. "Yeah, there's two ships gone already. I've logged the *Empire Union*, but the *Melrose Abbey* didn't get time to send the distress call. The mate said her back was broken and the bridge on fire. She went quickly."

He nods his head. "I didn't see the *Union*, but I saw the *Abbey*. The explosion lit up the sky as I came out on deck. It looked as if all hell had been let loose. The mate's right, the bridge was on fire. The radio room must have gone up in flames."

John sits in the operator's chair, his shoulders sag and he mutters, "The sparks must have had it. Burnt to death. Christ Almighty, what a death."

I feel a little sick at the thought, and hastily ask, "Did you see the U-boat?"

His face turns towards me. Somehow he looks paler than usual and the adolescent pimples stand out. "Yes, I saw the bloody thing. It was on the surface between the columns. I saw it clearly. With all those snowflakes, it was like day." He rubs his eyes. "She tore out ahead of the convoy and dived before the escort could get at her."

John pauses as he signs on the log and reaches for the earphones. His voice faltering, he continues, "There must be over a hundred men out there in the water. Poor sods. God, it'll be difficult searching for survivors at night and in these seas."

I nod wearily and reply, "Yeah, I can believe that. The rescue ship can't search long. If they're not found within two or three hours, they're goners."

Stretching, I move towards the cabin door, "Well, I've had enough. It's all yours now. Everything's logged, including the messages from Rugby Radio. Take care."

I'm tired; my brain, clogged with the incessant din of static and Morse to which it has been subjected over the last four hours, demands sleep. But an inner more desperate need to get out of this metal coffin and into the fresh air where I can see the threat and danger drives me to the outer deck. Dressed for the weather in a thick bridgecoat from my cabin, I wander out on deck. The cold air on my face clears my head and settles my nerves as I listen to the crash of the waves and the steady reassuring thud of the ship's propeller. *Yeah*, I think, *it's good to get out of there for awhile.*

As my eyes become accustomed to the darkness, I can see the dim outlines of the nearby ships. It looks peaceful, but looks are deceptive. I am no stranger to war; the London blitz saw to that. There's a tension in my guts, because this form of warfare is new to me. I've felt fear before, and now I feel it again, only this is fear of the unknown.

My mind goes back two years to that late summer night in 1940 and the fear I felt as an uncomprehending boy of fifteen. The roar of an approaching German bomber and the whine and shriek of the falling bombs are burned into my brain. The old farmhouse where I lived was close to an anti-aircraft battery situated in the green belt around London. I remember screaming, "Down, Dad, down!" but I doubt whether he heard me.

I dropped to the ground, hugging the earth and, looking through the palings of the fence, I watched the long string of exploding bombs coming directly towards me. I lay there trembling, my fingers working spasmodically, instinctively trying to dig myself into the ground. The earth shook with each explosion as the bombs crept closer. The shrieks from the fluted fins reached a crescendo as the flashes and thunder of the explosions came up and over a ridge in the fields, to the bluebell wood at the back of the house where my sister and I used to play.

As the last high explosive bomb tore into the wood not a hundred yards from the house, I hid my head, holding my breath. Almost simultaneously, I heard the short high-pitched shriek of a large oil-filled incendiary bomb which exploded and burst into a blazing furnace less than fifty yards away. I lifted my head, relief flooding my body, thankful to be alive, and stared at the scene. On the footpath in the back field was a ball of fire. Smoke and flames from the burning oil scorched a line of nearby elm trees. The flames, twenty feet high or more, illuminated the trees as dark black smoke drifted in the night breeze.

Yes, I think, *I know about that type of war*. I had watched other bombing raids from the shelter of the farmhouse porch, and listened to the whirr and clatter of shrapnel hitting the trees and the roof of the farmhouse. In the daylight raids, black puffs of bursting shells from anti-aircraft guns spattered the sky. At night the shells burst like fireworks, covering the night sky with bright flashes of light. For me, it was a world of excitement, a world of bombs and craters, old ruins and souvenir shrapnel, bits of bombs, but most of all, mine craters. These huge craters became a playground for my younger brothers and me.

Now, as I lean back against the ship's rail on the boat deck, war is no longer a game. *There are men out there trying to kill me. They've already got some of us. Who'll be the next?* My mind is too active for sleep. I have to calm my thoughts; the last four hours listening and guessing at the sounds of war have unsettled me. The wintry air sends shivers through my body. Instinctively I shrug my shoulders further into my bridgecoat and pull the high collar up. I find a sheltered spot on the deck under the bridge to shield me from the biting wind and rain squalls.

Astern, only the night stares back, unblinking, hiding the horror enacted in the rain-lashed seas beneath its dark blanket. Bitterly, I think of the useless waste of lives. Men are still struggling in the sea, but many are already dead. Our convoy has one ship designated as a rescue vessel, the *HMS Toward*; however, she cannot afford to venture far outside the convoy defensive zone without an escort. That's small comfort to those left behind struggling for their lives in the sea.

The back of my neck prickles in horror, and my hands grip the ship's rail as I drop instinctively into a crouch. The night erupts in a blinding flash which lights the sky towards the starboard rear of the convoy. A heavy explosion rocks the *Scottish Heather*; an instant later the ship's klaxon sounds shrilly. *Christ, they're attacking again!* It's about ten past four, barely ten minutes since I came off watch.

Snowflakes rocket up into the sky in confusion, lighting the convoy. A U-boat on the starboard side of the convoy has renewed the attack. She's penetrated the convoy unseen and torpedoed the last ship in the outer port column, the *Soekaboemi*, a Dutch ship. Bathed in the light of the snowflakes, the U-boat is heading at full speed across the columns, running trimmed down to her conning tower.

Standing transfixed, I grip the ship's rail hard, my hair bristling. The merchant ships are like a flock of sheep, only we don't scatter and run. The ships do not deviate but follow each other, one by one, like sheep to the slaughter.

The silhouette of the U-boat's conning tower is easily distinguishable as she tears along between the columns towards the leading ships. Crossing the columns, the U-boat targets the leading ship in column eight, and launches a torpedo. The *King Edward*, a large vessel of approximately five thousand gross registered tons shudders and heels over with the force of the explosion. She rights herself momentarily and then her boilers blow up in a second explosion. The flash reflects off the low overcast clouds in the sky.

The U-boat spins like a top and heads for the open sea. She has left the safety of the convoy. Freed from the fear of hitting each other with their fire, the gunners on the merchant ships open up on her. The staccato sound of machine guns and Oerlikons add to the confusion as ships and escort vessels fire at the U-boat, which is racing at full speed ahead of the convoy. Her conning tower presents an elusive target and she dives, disappearing beneath the surface. Horrified, I stare at the bewildering scene, my skin prickling at the sight of the final agonies of the vessels and the eerie light of drifting snowflakes floating down on their tiny parachutes.

We do not pause, cannot pause, to help survivors. The shattered ships, their speed suddenly reduced from eight knots by the inrush of sea through the gaping holes in their hulls, lie helpless, burning in the night, as they fall astern of us. The *Soekaboemi* is no longer in view; she's gone. The *King Edward* is close, broken in two, the jagged halves crumbling together. She goes quickly as I watch. In roughly three minutes she disappears beneath the waves, leaving indistinct blobs of wreckage and men struggling in the sea. The survivors swim or row, trying frantically to get away from the vortex caused by her sinking. They know that the wreckage which the whirlpool has sucked down will resurface with devastating results to anything and anyone in its path.

I hold my breath as I watch these sights. "Poor devils," I mutter to myself. I fight back the tears and swallow hard to lessen the lump in my throat. I know that men, both young and old, have died terribly. Some had burns and shattered bodies, and others drowned in the icy waters of the North Atlantic. *Oh, God,* I think to myself, *what a death.* I shudder as if I can already feel the numbing cold of the water on my body. *How long can a man survive in these waters? Probably five minutes, unless he can get on a lifeboat or raft.* Four ships have gone. The rescue ship, *Toward*, will be busy searching for survivors.

The need for sleep overcomes my desire to stay on deck. I stumble to the heavy metal weather door on the boat deck, and step inside, pushing aside the heavy blackout curtains. My cabin is close to the radio room and I can hear the faint rhythmic sound of Morse as I enter my cabin. Exhausted, I slip out of my bridgecoat and uniform jacket, remove my shoes, and fall asleep on the top of my bunk.

The ship's bell tolling the change of watch wakes me. It's eight thirty and the steward is banging on the door. Breakfast is being served in the wardroom. The cold water splashing on my face as I wash revives me, but my legs feel shaky as I make my way to the wardroom. The smell of bacon and eggs is nauseating and my stomach revolts. Swaying against the roll of the ship, I leave hurriedly to walk up the passageway to the deck.

I'm on watch again at noon. That's three hours away. My head is a bit woozy, not so much with seasickness, but because the cold I caught a few days earlier is getting worse. *Maybe the outside air will clear my head.* As I step out of the passageway onto the open deck, an icy wind bites my face. I pull the cowl of my duffel coat close over my head against the chill, and hurry to the sheltering bulkhead of the boat deck.

The images of last night's attack are still fresh in my mind. I'm drawn to the deck like a magnet. Instead of lying down in my bunk and resting my aching body, I am here in the shelter of the bulkhead looking at the grey cold sea. It stares back at me, cruel, heartless, and unforgiving. The convoy, battered by the initial attack, ploughs through moderate to heavy seas under an overcast sky and rain squalls sweeping across the ocean. The search for survivors has been called off. The rescue ship *Toward* is back on station.

Perhaps it's fear that keeps me from my cabin, a claustrophobic fear of being trapped in there. I don't know. Maybe it's just the seasickness and fever. I'm beginning to sweat, then shiver; my body feels strange, my arms and legs are leaden. I didn't eat breakfast this morning, because the sight and smell of the food made me feel sick. Here, the cold air on my face keeps the sickness within me. Huddled in my duffel coat, I try to retain the warmth in my unheeding, chilled body.

"The black pit!" I mutter to myself. "The second mate was right; it's well named." Scraps of thought flicker through my mind. *Where's my lifeboat? We've never had a lifeboat drill since I came aboard and that's two weeks ago*. I look aft, towards the stern. A catwalk stretches across the tanks from the main accommodation to the stern. The outslung lifeboats hang on their davits, ready for lowering in case of trouble. My gaze takes in the intricate valve and pipe system covering the tank deck below the catwalk. It is a stark reminder of the deadly cargo of fuel oil carried in the belly of the ship. I shudder. *Christ*, I think, *if that bloody stuff caught fire I'd be roasted alive before I could get to the bloody stern!*

Two hours have passed, and I'm slumped against the bulkhead on the boat deck sunk in my thoughts. Partly sheltered by the overhanging starboard wing of the bridge, I'm in the lee of the prevailing weather. Odd swirls of rain occasionally drive in, but I hardly notice them. Only nature and its basic functions move me; the ship's head calls.

My cabin is warm as I enter and hang the duffel coat in the cupboard. John, the second sparks, is on watch; I'll be relieving him soon, but now I must dash.

The head, or washroom, is one deck down and there's an axe on the bulkhead of the toilet. I smile as I see it. Strange that its presence has not registered with me before. Perhaps I have taken it for granted; that is, until the junior engineer told me about the chief engineer.

My thoughts go back to Christmas dinner. The wardroom was gaily decorated with long streamers of coloured papers, large bells and holly; when I entered, I couldn't help but notice that several of the ship's older officers were wearing life jackets. It looked odd against the background of festivities.

As I sat down next to him at the table, a young junior officer said in a low voice, "Look, Sparks, the chief engineer's got his life jacket on! I hear he even takes it and an axe to the ship's head with him."

I laughed, then replied, "Well, maybe he has his reasons; he's not the only one. But why the axe?"

Leaning towards me, he confided: "From what I hear, he's been torpedoed several times. On one ship he was caught on the john, and the door jammed. The metal frame twisted with the force of the explosion. With an axe you can smash the door down. They're standard equipment in the ship's head, but the chief carries one with him just for peace of mind, I guess."

The incident seemed funny to me at the time, but now, after last night, it isn't funny any more.

Noon. The ship's bell sounding the change of the watch disturbs my thoughts, and I hurry to the radio room; it's time to take over the watch. The musty damp odour from the steam heater in the radio cabin catches in my throat as I sit down in the operating chair, and reach for the headphones. *Ohh, no! I can feel the queasiness in my stomach growing.* My eyes glance apprehensively at the red can; the can and I are inseparable companions. Motion sickness and fever drain me of energy, reaching a peak of wretchedness as I throw up in the can. I've had no food since dinner last night and I bring up green bile.

Today I feel bad, although the motion of the ship is not violent. It must be the fever burning in my body; I have hot and cold spasms. Momentarily my stomach settles after a bout of sickness. Once again I begin to function normally, until the cycle repeats itself. The small electric fan provides some respite as I switch it on, and feel the warm air play on my face.

The crackle of static, interwoven with the stream of incoming Morse signals, drills into my brain. I listen intently to the coded radio bulletins. There are more warnings from the Admiralty of increased U-boat radio activity as latecomers join those already here. They are like packs of wild dogs attacking and pulling down their prey.

The messages are piling up, and I call the bridge on the intercom system. The second mate answers, then hurries down to collect them. He looks at me as he stands in the doorway. The smell of vomit lingers in the room as I sit in the operating chair, the messages in my hand. "You all right, son?"

I manage a sickly grin. "I guess I'll survive, only the bloody ship's bouncing too much for me. I think I've got a bit of a fever. Probably coming down with the flu. My legs and arms feel like lead." I hand him the messages and continue, "These should interest you. Looks like more trouble."

He takes the messages, glances at them and nods. "You'd better take something for that cold; you don't look well at all. Anyway, I'd better get these to the bridge right away; the Old Man will be interested in them. We're scheduled to refuel one of the escort, the *Chilliwack*, later this afternoon. We'll be dropping behind the convoy again about sixteen hundred hours." Turning, he disappears along the passageway.

Somehow the four-hour watch passes. The *Chilliwack* has arrived and, judging by the noise of the winches, the deck crew are preparing to haul the hawsers across the gap between the two vessels. I look up with relief as the chief comes in to take over the watch. There is concern in his voice as he sees my flushed, dishevelled appearance, and smells the sweet sickly odour of vomit.

"My God," he says, "you look awful. How do you feel?"

Wearily stretching my cramped legs and arms, I look at him through blurred eyes. "Bloody awful, Chief." I pause a moment, then go on. "My legs feel like lead and my head hurts like hell. I think I must have caught the flu or something."

The touch of the chief's hand against my forehead is cool. "Yes," he replies, "you've got quite a temperature. You'd better turn into your bunk, and I'll get the steward to bring you some aspirin. It will help get the fever down." He hesitates. "Have a good night's sleep and we'll see how you feel in the morning. Now, don't worry. John and I will cover your watch tonight."

On unsteady feet I leave the radio room and make my way down the passage to my cabin. Drawn by the commotion on deck and the rattle and squeal of steam winches, I open the heavy outside weather door leading onto the boat deck. The Canadian corvette, *HMCS Chilliwack*, lies off our starboard beam. The tanker's bulk dwarfs her as she rides the rough seas, rolling and pitching at reduced speed to match that of the tanker. We're going slowly, about four knots, and heavy steel stranded hawsers are stretched out taut between the bows and sterns of the two ships. As I watch, the pipeline, supported by floats, is hauled across the gap between the two ships to the crew waiting on the corvette. I hear the bosun's voice as he yells to a crewman, "Careful, Nobby, watch that bloody line."

My curiosity satisfied, I head back to my cabin. The fresh air has helped to partly clear the headache and sickness, but my body feels leaden as the fever burns in me. A smile flits across my face as I see that the steward has left a bottle of aspirin on the bunk side table. It, and the sight of the clean white sheets and blankets on my bunk, drive all thoughts of the outside world, and the dangers lurking below the surface of the sea, from my mind. We are alone fifteen miles behind the convoy, refuelling the corvette, and dusk is falling.

Torpedoed

THE EXPLOSION shatters my cabin, precipitating me from a deep, drugged sleep to a state of bewildered consciousness. My mind is befuddled; it cannot comprehend the ear-splitting roar and blast. Debris and dust fill the air, raining down on my body in the bunk. A heavy weight pins me down. My hands, trapped beneath the entangling sheets and blankets, fight instinctively to get free.

I awaken to a nightmare and shake my head in panic, but the ringing in my ears will not stop. There is a strong pungent odour of explosives. My body is forced against the side of the bunk as the ship heels over to port. Realization comes to my sleep-clogged brain: this is real; it's not a dream. My heart pounds and my stomach muscles tighten involuntarily. *We've been torpedoed!*

Everything is happening in slow motion. The ship is shuddering, listing hard over. *I've got to get out. She may go at any moment!* Panic galvanizes my body and my throat tightens with fear as I struggle to free myself. *I can't see!* The cabin is dark because the explosion has killed the lights. I can hear the screech of tearing, twisting metal and the roar of rushing water; it terrifies me. *The tanks must be ruptured!*

My heart is racing as despair surges through my body. In desperation, I pull my hands free of the blankets and touch jagged splintered wood and a smooth metal knob. It's the cabin door, blown on top of me by the force of the explosion. It falls with a crash as I push it off and tumble out of the bunk. The roll of the ship has stabilized; she's not going over. Disoriented and barefoot, my head aching with fever, I scramble in the darkness, heedless of the rubble. The horror of the sea pouring into the bowels of the ship and the thought of drowning here in my cabin excludes all other thoughts as I concentrate on survival.

Outlined against the faint red glow of the emergency lighting in the passageway is the rectangular opening where the door used to be. The passage is deserted. I run to the heavy outer door and thrust it open. Clad only in thin linen pyjamas, I step out onto the boat deck and into another world, a world of confusion.

Flares illuminate the night; they hover high in the sky, then drift down, bathing the sea with their light. Streams of tracer bullets, aimed at some distant target, scream across the sea. I catch my breath—the bullets are very close. Then, realizing they come from the *Chilliwack*, I exhale slowly. The corvette is going into the attack, firing every gun she can bring to bear on the U-boat as it disappears into the night.

I glance towards the bridge. Another flare detonates, hurling its container into the sky. It bursts, then it, too, floats slowly down. The frightening roar of water is louder now, mingling with the confused shouting of men. I hear the ominous sound of tortured metal plates grinding against each other and tearing off as the ship gradually loses way in the heavy seas.

Smoke billows from small smouldering fires which spread an eerie light over the ship. *Fire*! My heart pounds. If the fuel oil catches fire, it will explode and the whole ship will disintegrate in one huge fireball. *Oh, Christ Almighty, I don't want to die, not burnt to death!* I bite my lip to control myself; sanity returns and I feel calmer. The smoke is from smouldering debris and the ship is tilting more slowly now.

The initial shock is wearing off and I shudder, shivering uncontrollably. I am suddenly aware that it is freezing cold. I can't feel the wooden deck beneath my bare feet. *I must get some clothing and my life jacket.* Sick, and not thinking, I've made a near fatal mistake and broken the first rule of survival: in a war zone, sleep in your clothes, or perish of hypothermia in the boats or sea.

Summoning up my courage, I dash for the passageway door. If the ship goes now, then I will go with her, but if I stay, I'll die in the icy water. I've no option.

Ramming down the weather door handle in my haste, I yank the door open and enter the passage. Fixed high on the bulkhead, the small red light, powered by the ship's emergency batteries, casts a soft glow in the darkened interior. Finding the opening to my cabin, I enter and search amongst the wreckage. The cabin is a shambles.

Fighting off the claustrophobic fear of being trapped and drowned, I grope in the darkness for my clothes. Bending down to crawl on hands and knees in the wreckage, I feel the blood rush to my head and a wave of sickness comes over me. But the thought of drowning forces back the sickness in my throat. "For Christ's sake, help me," I sob aloud, fearful of the ship taking another disastrous tilt. I touch the thick cloth of my bridgecoat on the floor of the shattered cupboard. Time seems endless as I search for my life jacket. "Aaahh," I cry out in relief when my scrabbling fingers feel the bulky kapok. Underneath the life jacket, I find my shoes.

Stepping into the dimly lit passageway with my clothing, I slip my frozen feet into my shoes. As I don my bridgecoat and life jacket, I feel a measure of self confidence returning. *I must find the chief.* Instinctively, my steps turn towards the radio room farther down the passage. I will find him or John there.

The room is empty! The emergency light spreads a soft glow over the wreckage. The main radio transmitting equipment lies smashed beyond use. Undecided, my eyes take in the old spark gap transmitter. It's robust and reliable. Somehow the sight of it calms me, because the transmitter appears serviceable. I hesitate, undecided what to do. Questions race through my mind. I glance at the clock still screwed to the wall. The glass is smashed but the hands show a little after eight o'clock; it's the chief's watch. *Yes,* I think, *he'll have already sent the distress call. But where is he? And where's John?*

There's no point in staying here. The chief's probably on the bridge with the captain. I run down the passageway and out to the boat deck. As I make my way to the bridge, I bump into the chief.

"Chief," my words come in short bursts. "What's going on? How badly have we been hit?" We walk together across the boat deck towards the radio cabin.

"I've just been on the bridge with the other officers. The Old Man's given orders to abandon ship," he says. "We've been hit just for'ard of the bridge on the starboard side, blowing several tanks. We'll have to get the emergency radio equipment into a lifeboat right away. If another torpedo hits anywhere near that crude oil, we've had it."

John's waiting for us by the radio room. "What a bloody mess," mutters the chief. "We've got to find the lifeboat transmitter." He gestures, pointing with his hand. "It should be in that locker."

With the aid of a flashlight, we rummage through the debris. Dragging the transmitter in its waterproof case from the locker, John and I carry it to the boat deck.

Relief sweeps over me as I feel the cold air against my face. At least, now we have a chance to jump clear if the ship suddenly slides beneath the waves. My head throbs as the suppressed fever surfaces, making me sweat and shiver alternately. My legs and arms feel weak from the effort as we rest the case momentarily on the deck while John talks to the chief. Their voices are faint, remote, as I sit on the box fighting back the vomit rising in my throat.

John's voice cuts through my thoughts. "Come on, Gord," he says, "we've got to get it down to the next deck."

Lightheaded, I bend and grasp one end of the case. Together we carry it across the deck to the companion ladder. It is dark; neither moon nor stars light our way, just the dim, dancing beam of the flashlight held by Dick, our chief. As we make our way around vague outlines of the ship's superstructure, I stumble, hitting my toe on a deck bolt.

"Careful," John's voice is strained. "We'd better not drop this bloody thing."

We carry the case down the ladder to the main deck, and prop it up against the ship's guard rail. Scrambling nets and rope ladders are already in place over the side, and men are climbing down them to board the lifeboats. The top rail is cold and wet against my hands as I grip it. Looking over the side, I see the shadowy outlines of two lifeboats, numbers one and two, tossing in the sea below.

The indistinct shapes of the lifeboat crew move about in the boats, their flashlights swaying as they work in the dark. Escaping crude oil damps down the waves, making their task a little easier. Their voices float up. "Get that fucking rope tied, Nobby," says one. Then, another adds, "For fuck's sake man, use the bloody boathook to fend off the goddamned ship's side, or you'll have us all in the bloody drink."

I ask John which lifeboat is his. He points to one of them and says, "Number two, just aft of the for'ard lifeboat there." He turns to the chief. "Which lifeboat do you want the radio in, Dick?"

"Aaahh, what the hell, John." He pauses, then adds, "Put it into the nearest, but you'll have to go with it."

For the next few minutes John and I are busy lashing one end of a long rope around the transmitter case. The ship is dead in the sea, and we are all too aware of the impending danger. *It'll only take one torpedo to finish us.* Working with John takes my mind off these thoughts as, between us, we balance the case on the ship's rail.

Holding it there, I watch John scrambling down the Jacob's ladder to the lifeboat below. His dim form, disappearing down the side of the ship, is outlined against the shaded flashlights of the boat crew. He hesitates at the bottom, waiting for the boat to rise up on a wave, then jumps onto the broad planking of the lifeboat seats.

Standing in the well of the lifeboat, John looks up at me. I can make out his pale, indistinct white face and see his mouth opening and closing. He must be yelling at me, but his voice is drowned by the roar of water pouring into the ship's hull, and the shouts and curses of jostling men clawing their way down scrambling nets and rope ladders. He gestures with his hands, and I ease the bulky case over the side.

The rope takes the strain. Slowly, hand over hand, I lower the case until it dangles a few feet above the boat. Despite the oil, the waves thrust the lifeboat violently inwards and upwards as it rises and falls in the sea. A crewman with a boathook fends the lifeboat from the ship's side, while John and a seaman struggle to catch the swaying case. They haul it into the boat as I pay out the rope. The muscles in my neck relax momentarily. The transmitter is safely aboard now and stowed under the rear seat.

I turn and walk across the deck towards the short ladder leading to the catwalk. The ladder is only a few feet away, but those steps are critical. My fingers grip the hand rail, grip it hard, as I stand there at the bottom of the ladder. Something makes me pause. The pungent smell left by the explosion and the smoke masking the catwalk strike dread into my heart.

It seems an eternity, but it is only seconds that I stand there. The black background of the night frames the smoke and the iron railings on each side of the narrow raised catwalk. In my mind, the flickering lights enhance its menace as the catwalk disappears into the smoke. It stretches across the huge tanks leading aft. The scene is unreal, and magnified by my fevered brain. I am looking into a great frightening void, into which the catwalk leads like a giant staircase.

My thoughts are torn. I want to go to the second mate and my lifeboat, but an instinctive fear of crossing that void alone holds me back. If only the second mate were here. I would gladly go with him. His voice comes back to me, "Listen, son. If there's trouble, stick close to me. Our lifeboat's on the aft deck, number three on the starboard side."

Confused, I am hesitant to walk across the catwalk. If the U-boat puts another torpedo into us, those tanks will erupt in a fireball beneath me. Bewildered and uncertain, I am ashamed of my feelings and shirk the responsibility of making my own decision. I turn to my chief and appeal to him for direction.

"Chief," I say, unable to suppress the tremor in my voice, "my lifeboat's aft; it's the second mate's lifeboat." My throat is dry as I point to the catwalk. "I . . . I don't like the look of that. What d'you think? Should I go?"

The chief glances at the catwalk. "I don't see that it matters, Gord," he says. "Nobody's in their right boat, anyway. Go with John; he'll need help with the transmitter. But you'd better hurry. I'm going in the captain's boat."

The chief turns and hurries away across the deck. I watch him disappear, a dark shadow in the night. There is no time for second thoughts. I climb over the bulwark and onto the Jacob's ladder. My feet instinctively find the wooden rungs as I grip the rough rope sides with my hands.

The ladder sways drunkenly. Men on the ladder below and above me are indistinct shapes who tread on one another and curse. They are anxious men; men intent on survival. Their panic-stricken voices, partly lost in the babble of noise and humanity, call desperately to their shipmates. "Hurry or another fucking torpedo'll get us."

Clinging to the ladder, the dark broken only by the flicker of the shielded flashlights, I wait for the man below me to jump. Watching the turmoil, I feel the tension rising in my stomach. The smell of oil is thick and penetrating. I lick my lips and try to swallow the saliva to assuage the dryness in my throat.

The lifeboat is held against the side by ropes fastened to the ship's rails. I can see the outlines of the ropes and tackle used for lowering the lifeboat from its davits. They have been cast off and hang down loose. Scrambling nets line the sides of the ship. Men cling to them like flies, as they make their way to the waiting lifeboats.

The lifeboat crew use canvas covered rope, kapok buffers and boathooks to keep the boat from being dashed against the vessel's side and overturning. I feel the ladder jerk and sway. My turn to jump has come. I edge my way down and hang on to the last rung of the ladder. Then, as the lifeboat rises up on the crest of a wave, I jump into the vague shape of the boat below.

Landing awkwardly on the slippery wooden planks, I sprawl forward, one foot in the well, my body cushioned by crew members already seated in the lifeboat. A hand grips my shoulder as I fall. "Steady, man, steady," someone says. "You'll break your bloody leg." The breath is knocked out of me. Scrambling to my feet, I mutter apologies.

"That you, Gord?" I recognize John's voice and make my way towards him through the seated figures. He shifts to one side, making room for me beside him in the stern.

"The chief told me to join you, and help with the transmitter. He said that nobody was in their right boat anyway."

"Yeah, well that's for sure, I can do with the help." He grunts in disgust. "No wonder there's confusion; we've never had a bloody boat drill. Did you ever see your lifeboat?"

"No, I've not been aft on this ship. All I know is that it's number three on the starboard side."

We lapse into silence as we wait for latecomers and the order to cast off. Preoccupied with our own thoughts, we watch the activity around us. Hardly aware of each other, we sit there, tensely waiting, expecting the sudden savage blast of an exploding torpedo. There is a strange sensation in my skin; it is a prickling, cringing sensation as my muscles tense for the inevitable. My throat is still dry. I lick my lips again and try to swallow, but it hurts. The men around me are calling out, "Come on! For fuck sakes, hurry, man. Let's go!" The order to cast off is long in coming.

The second mate's words come back to haunt me as I sit in the lifeboat, wondering where he is. I feel I have forsaken him. He's the only man who has shown kindness and interest in me since Dad died. "Stick to me, son," he had said. I feel tears of humiliation not far away. *Have the years of unquestioning obedience to my teachers, the nuns and the brothers, and to my mother, not to mention the years of strict discipline as a cadet in the Air Defence Corps, deprived me of the ability to make my own decisions?*

There's a lump in my throat and I swallow hard, looking around as I desperately try to forget the feeling of guilt. I watch the last shadowy forms scurrying down the nets and rope ladders. An indescribable ache settles in my throat as I sit frozen to the seat, my body tensed for death, and I half sob in desperation as I urge them on in my mind. *Hurry, for Christ's sake, hurry.*

The ship's hull, large and monstrous, looms over the tiny boat. *Is it my imagination, or is the tanker tilting more and more over us?* I hold my breath. *Will we never get away?* Time stands still as we wait. Finally, the last man is aboard and the order to cast off comes. I feel the tension draining from my body. The ends of the holding lines fall into the oil-covered sea to dangle limply down the ship's side. But the relief is just an illusion. There is no easy path. Almost at once we run into more trouble.

Manning their oars, the lifeboat crew pull desperately as they strive to edge the lifeboat away from the side of the ship. Seated side by side, their muscles strain as they dip their oars into the sea. They redouble their efforts, but it makes little difference. It's as if a giant hand holds us against the oil-grimed side of the ship. Dashed backwards by the oncoming waves, we are in danger of being swamped and overturned.

The coxswain suddenly realizes the cause. His angry frustrated yell splits the tension in the night air. "What the fucking hell?" he shouts as he fights with the tiller. "The bloody cradle's still under the fucking boat. Cast the bloody thing off, for Christ's sake."

There is a flurry of movement as crew members frantically release the two cradles encircling the underside of the boat. As they sink into the depths, the lifeboat responds to oar and tiller. We begin to move slowly, inching our way diagonally from the side of the ship, heading for the open sea.

We gain a little leeway and move a few yards from the stricken vessel. In the confusion, the gigantic hole torn by the torpedo in the side of the tanker is forgotten. Our stern jerks round violently as the strong drawing force of the water pouring into the ship takes hold of the small boat. Like an enveloping great mouth, that black hole is frightening. Our boat crew fight for their lives and ours, as they bend with renewed strength at the oars.

Fighting against the seas and racing water that threatens to engulf us, the oarsmen strain desperately to haul the lifeboat out of the current. Gradually, at a drunken angle, we slide past the hole and the ship's bow to the sea beyond.

Night of Terror

THE SHIP'S BOW is astern now. The prickling sensation in my skin gradually subsides as the threat of annihilation by an exploding torpedo diminishes. We are in the open sea, but still near enough to be dragged down in the vortex, or killed by debris shooting up from the depths. I watch with anxious eyes as the ship fades from view in the darkness. Urgency seems to lend strength to the oarsmen, because the instinct of self preservation is strong.

The coxswain in command of the lifeboat issues an order to activate the life jacket lights. "Mates," he says, "if you want to live and be picked up, switch on your bloody life jacket bulbs. If you've forgotten how to do it, there's a small lead on the jacket. Stick the bloody thing into the battery. Got that?"

"Aye, Cox," mutters a voice. My numb fingers fumble with the lead as I plug it into the small battery in the neck of the life jacket. One by one, the little red lights spring to life. The effect is grotesque, creating an illusion of disembodied heads, red half heads, bobbing and weaving in disjointed lines against the backdrop of the night. One side of every face is bathed in red, while the other side is in the shadows. A shiver runs through my body. It's like a nightmare; these are ghouls whose expressions change in animated conversation.

We sit huddled together for protection from the wind and spray, while the waves break and divide round the climbing bow, then fall away as we slide down into the trough. The coxswain, his figure but a dim form in the stern, struggles with the tiller, concentrating on keeping the boat's head into the oncoming seas. I look away, leaning back in my seat against the gunwale, and listen to the sounds of the sea, the slap of waves and the murmur of men's voices. My eyelids feel heavy and I shiver with the fever eating into my body. The rhythmic creak of the oarlocks and splash of dipping oars come faintly to my ears.

The voice of the coxswain breaks the tension as he addresses the rowers. "Okay, mates," he says, "that's about it. We're far enough away now, if the old bitch goes." He pauses, then adds, "Just keep the bow headed into the waves while we get our bearings on the other boats."

He's barely finished speaking when someone shouts, "Fucking hell, we're sinking! Start bailing."

Sea water has crept over the wooden grating covering the keel or bilge well in the bottom of the boat. The bilge is full of water which is now creeping up our legs as we bail. Frantically fifteen pairs of unorganized hands dip into the water. We use anything we can find, hats or cupped hands, but we cannot hold our own. Momentarily the boat loses way as the rowers abandon their task to bail.

We are slewing round broadside to the waves as the tiller falls slack in the coxswain's hands. His voice screams orders. "Pull on the oars for fuck's sake, or we'll capsize in these waves, you stupid bastards. The rest of you, bail. Nobby, get someone to help you try and plug that goddamned leak with clothing, anything, for Christ's sake."

The oars dig in the sea again as the boat pitches and rolls. She rights herself and heads into the oncoming seas. Nobby's voice, half sobbing with emotion, cuts across the babble of frantic voices. "S'no use, Cox. Too many bloody holes, man. We can't get at 'em. The fucking grating's in the way. She must be riddled with shrapnel!"

"Shit! Leave them then and bail." The coxswain swears in frustration. "Christ all bloody mighty. We've got to get to the nearest bloody lifeboat, or we'll be swamped."

We need no urging as we renew our efforts, both rowers and bailers. The coxswain pushes hard against the tiller handle and steers towards the nearest row of bobbing red lights in the darkness.

The lifeboat settles deeper in the water. Our situation is getting desperate. My hands are icy and numb as I try to bail with my cupped hands. Crouched in the well of the boat I feel the bitter cold of the water creeping up to my knees. My heart is in my mouth and I half sob with the cold.

Giving up my position in the well, I scramble back on to the wooden planking adjoining the gunwales of the boat. I tuck my legs, frozen from immersion in the sea water, under me as I try to lift myself clear of the freezing water, and bail at the same time. It is but a momentary respite. The water drains from my shoes as I sit in my sodden pyjamas, my bridgecoat covering my shoulders and upper body. There is no way to escape the incoming sea water. If we can't make it to one of the other lifeboats, we will soon have to take to the sea.

I can't think clearly. Frightened, I try to pierce the blackness looking for little rows of red lights. We top a wave and I see them lying diagonally below us, a short distance away in the trough of a wave. A cresting sea catches us beam on and we lose sight of the other lifeboat as water crashes over our gunwale. It is only a matter of minutes now before we are swamped; our freeboard is perilously low. The lifeboat appears again out of the darkness. It is the bosun's boat; I recognize his voice as he replies to our plea to come aboard.

"No, mates," he yells back. "You're too many for us; we haven't the room. Make for one of the other boats. You'll swamp both boats if you try to board."

Stunned, my heart sinks and I catch my breath. *No, no,* my mind cries out, but not a sound passes my lips. *You can't leave us here to die*. Mechanically, my frozen hands dip into the water in the well of the boat; they form a cup to throw what little I can hold over the side. It is a hopeless task. I am beginning to give up hope, hope in humanity and my fellow men.

"Goddamned fucking son of a bitch," yells one of our crew. We turn away, rowing and bailing desperately to get near another lifeboat. The pitching tossing lights of a third boat appear directly ahead as we leave the bosun's boat behind.

Guided by the glow of the red life jacket bulbs, we slowly draw nearer. One thought occupies my mind. *Will we be turned away again?* If we are, I'm afraid that I won't last long in the water. I know that because my legs are already frozen. I can't feel the wet of my pyjamas against them but I can feel it against my belly. I shudder to think of the water closing over my head. *Christ, Dad, help me!* I thrust the thought from my head and frantically bail.

As we pull nearer, our coxswain hails the boat. His voice, competing with the roar of the sea and wind, carries across the open space between the boats as we toss, one moment on the crests of the waves and the next in the troughs. "We're sinking. Christ Almighty, man, let us aboard. We're fucking sinking," he says. "Our bloody boat's holed by shrapnel, and we're taking water fast."

I hear the captain's voice with a feeling of relief. The north country accent is reassuring. "Aye," he replies without hesitation, "we'll make room for you. Have you the emergency transmitter?"

"It's with us, Captain." John answers.

"Good. First pass it over; then you can all come aboard. We're going to need that radio."

"Lousy bastard," I mutter under my breath as I kneel in the well, helping John to haul the case out from under the seat. "He thinks more about the bloody radio than his crew. We could drown in the goddamned sea, for all he cares."

A fleeting vision of the old farmhouse, the log fire in the large grate in the living room hovers in my mind. *Will I ever see the old house again and my brothers and sister, or tread the forest path?* My mind snaps back to the present as the case comes free, and I'm back in the real world, the icy cold world of the North Atlantic.

The boats touch with a grinding crunch as we bodily thrust the emergency transmitter into the captain's boat. Water, glittering in the light of a flashlight, pours out of its case. A voice proclaims in disgust, "Jesus Christ, the fucking thing's supposed to be waterproof. Fat lot of good that is."

Right now, I can't care less about the bloody radio. Our lifeboat is sinking; only the buoyancy tanks and constant bailing keep it afloat. We're awash almost to the gunwales and cold and wet from immersion in the sea. It is difficult keeping the lifeboats together. The darkness doesn't help; it just makes it more nightmarish as the boats clash and grind against each other in the rising and falling seas.

The weight of men on one side of the boat forces it under water, and the sea pours in. I'm standing in the well of our boat trying to grip the gunwale of the captain's boat to keep them together. It's almost an impossible task as the waves keep forcing the boats apart. The water's up to my thighs. My Christ, it's cold. Men are cursing and a voice louder than the rest growls, "Watch it, you fucking bastards, you'll overturn the bloody boat. Take it easy. One at a time."

The boats partly separate and in the confusion men are half dragged and hauled from the waterlogged boat, or the sea. I manage to scramble aboard as the boats rub together again and I flop, sopping wet, in the well of the captain's boat. I gasp for breath and shiver where I have dropped, huddling from the wind, which bites through my bridgecoat. My head is woozy as I rise shakily to my feet and slip off my sodden pyjama bottoms to wring them out before pulling them on again. I hurriedly sink back into the well, trying to rub the circulation back into my legs.

A brief head count by the captain and coxswain confirms that we are all accounted for. *Yeah*, I think, *thank Christ that's over, at least we're alive, although I don't know for how long*. The lifeboat, its freeboard lowered by the double load, has between twenty and twenty-five men aboard.

From where I crouch I hear the coxswain talking to the captain. "Aye, sir," he says. "It was shrapnel from the torpedo, Captain. The boat was rigged outboard for quick launching. Shrapnel must have riddled the bloody bottom. We didn't notice it until we pulled away from the ship; then it was too late."

The conversation spreads as the boat crew join in, confirming the coxswain's words. Men begin to rearrange themselves among their friends. I'm cramped in the boat well and, as I get up, my pyjamas, despite my efforts to squeeze the water out, still flap wet and cold against my torso. I move cautiously amongst the tangle of bodies, trying to find a place to sit. "Here, mate," a voice says and a hand grabs me by the shoulder. "For'ard a bit; there's space up there."

It's crowded but reasonably dry, except for the salt spray from the tops of wind-whipped waves. Thankfully, my body feels the hard planking beneath me as I half lie on the bow decking. My limbs ache and physical weariness sets in, but I cannot rest. My feet are frozen. I have no socks, and the sea water squelches in my shoes. The water drains from them as I take them off and massage my numb feet, trying to restore the circulation and warmth. My bridgecoat, although wet, at least breaks the force of the wind. A measure of security returns to my mind as I begin to take stock of my surroundings.

My eyes focus on the shadowy outlines of a little knot of men grouped in the stern of the lifeboat beside the coxswain. They are the deck officers. The captain is there. The murmur of voices around me draws my attention. I listen in silence to the talk, but I can't seem to think at all. It's like my brain has been put on hold.

The abandoned lifeboat is barely afloat, but we are towing it on the captain's orders. Waterlogged, it lies astern of us, sloughing drunkenly in the sea. I can hardly see it in the darkness. The towing rope pulls and tugs, stretching taut as the abandoned boat crests a wave, then slackens dramatically as the boat rushes down the sloping wave, racing towards us.

The captain notices it; raising his voice, he calls to one of the crew members. "Lengthen the bloody tow rope," he says. "That boat'll stove us in if it hits us."

He turns and addresses the silent men, sitting and lying together. "The boat's waterlogged. It won't support your weight if you have to get in it; but it's something to hang onto if we overturn. We'll keep it in tow for tonight, just in case."

The captain's words barely register in my mind, and I dimly hear a crewman mutter to his mate, "Aye, Fred, that makes sense. Though if this fucking boat overturns, we can say our prayers, man. That bloody sea's freezing. We won't last long."

The rope is hurriedly lengthened and the abandoned boat falls further astern. The oarsmen dip their oars into the water and pull back on the heavy blades, conserving their energy, while keeping the boat's bow headed into the waves. A nerve twitches in my face, and I tremble with fever, sweating one moment, then freezing the next. I bury my head in my arms while, in a delirium, I drift in and out of reality.

A sudden commotion in the stern of the boat disturbs my wandering thoughts, bringing me out of my lethargy. The low hum of whispering voices stops. A flashlight shines on an oar blade. One of the officers is tying something to it. I realize, with a start, that it is a flare.

The flare is ignited and held aloft on the oar. It bathes the lifeboat and the surrounding sea in a ruddy circle of light. We sit in the eerie red light, waiting tensely for the sound of engines from the rescue ship or escort. Only the splashing of waves against the boat's sides breaks the stillness.

I feel a sense of desertion as hope fades. The *Chilliwack* has left us, and we are too far behind the convoy for the rescue ship to attempt to find us. Disheartened, I watch the flare slowly die. The returning night takes on a deeper darkness which matches my despair.

Huddled with the others in the dark, I sink into bitter depression. An order from the captain is passed from man to man. A crewman shakes my shoulder, startling me from the stupor into which I have sunk. "Aye, lad, wake up," he says. "Switch your bloody light off. The escort seems to have gone and the Old Man don't want to take no chances with them U-boats machine gunning us. You know, the bastards have done that before." My fingers, numb with the cold, fumble with the plug on the life jacket collar. Disconnecting it, I pass the message on.

It's about three hours since we took to the boats. The captain, his voice raised, addresses the crew. "Men, we'll have to wait here till dawn. If the ship's still afloat, we'll board her and try to make port." He pauses a moment, then his voice, pitched lower, reflects his fatigue as he says, "I'm sorry, but rescue seems out of the question. We're too far behind the convoy. We'll just have to make the best of it. We are going to put out the sea anchor to hold the bow head-on into the sea so we don't get swamped. Is that understood?"

A murmur of assent follows his words, and two crewmen begin assembling the anchor, working quickly by the light of their flashlights in the well of the boat. Then, finished, they throw the anchor, shaped like an airfield windsock, overboard where it floats in the sea. The lifeboat, tethered to the sea anchor, swings freely, responding automatically to shifts in the oncoming seas. We settle down for the night, waiting for the dawn.

I hear a moan and turn to see where it comes from. A dim shape lies huddled close to me. The white of his shirt shows below the dark life jacket. I grip his arm. "You okay, man?" I ask.

He is so cold that he has trouble speaking. His voice is almost sobbing as he mumbles a reply. "I . . . I'm the . . . the galley boy. D'didn't have time to get clothes."

I remember seeing him in the galley. He's young and has not been long at sea. I put my arm around him. He's freezing cold. All he is wearing is a short-sleeved shirt and a pair of trousers. I unbutton my bridgecoat and wrap it round his trembling body and mine.

His voice is low, hardly audible, as he murmurs, "Thanks, mate."

We lie there, his wet hair just under my chin, holding each other, trying to get a little warmth into our bodies. Too exhausted and tired to speak, we just need the compassionate touch and warmth, as we huddle together against the violence of the weather and sea.

Shivering, then sweating, I retreat from reality, gazing unseeing into the night. It is a world peopled by memories and the dull aching throb in my head. My mind wanders among the ghosts of the past to my childhood. The dim shapes of the nuns swim indistinctly in the dark and trigger bitter memories. I feel anger stirring deep within me. My hands are tied behind my back and my buttocks and legs are stinging with pain. They are whipping me with their thick leather belts. Through my tears, I see the indistinct mass of children, my classmates and, yes, my two younger brothers. They watch as I stand on the raised dais. Sister Philamena is there; strange how I remember her name. A feeling of deep shame sweeps over me. *Why? why did they do this to an eleven-year-old boy?*

A leg muscle cramps. I stretch out my leg, and the pain eases as I shift restlessly under my bridgecoat. My thoughts turn to the Brothers of the Order of the Sacred Heart who ran St. Egbert's, the religious college I attended. The black and white robes of the nuns are replaced by the austere dark brown, ankle-length garb of the brothers. Cowls hang down the back and large red crosses are sewn to the front of their gowns. White clerical collars encircle their necks. There is no mistaking these brothers.

Brother Finbar's jeering face forms before my eyes. The sadistic bastard beats me with a three-foot wooden chalkboard compass, beats me on the buttocks until tears run down my face. He catches my tears on the end of the compass and displays them to the class to humiliate me. "Look," he says, "real tears . . . a crybaby! See the tears. Look at him!" I don't cry out; I won't give him that satisfaction, but I am learning to hate.

Cramping leg muscles again awaken me to reality. I try to stretch and massage my legs in the confined space where I lie on the wet boards. The galley boy moves slightly and clings to me.

"It's just leg cramps," I tell him as I rub the aching muscles to restore the circulation and ease the pain in my wet icy legs. There is no feeling in my feet. My shoes are wet and sodden, but at least they provide some protection.

My mind wanders unfocussed as I settle down and readjust the bridgecoat over us. My father's face, as he lay dead on the old-fashioned couch in my grandmother's house, comes back to haunt me. In my fevered mind, his face changes to that of the second mate. I wonder where he is.

Odd murmurs from men talking to one another in low tones carry to my ears as we huddle in the bridgecoat. Cascading icy cold water shatters my thoughts. I sit up with a start.

An errant wave hits the lifeboat's forward starboard quarter as we swing on the sea anchor. The boat lurches violently, its counter dipping, then rears upwards on the dark looming slope of the wave. I catch my breath and hold it for a terrifying moment. The boat rights itself, and I breathe easier as the wave passes under us.

There is a flurry of activity as crew members bail. Then, the crisis is past and the boat settles down to swing once more on the sea anchor. I close my eyes, shutting out the murmurs of others and the incessant slap of cresting waves as the boat rides them, pitching and tossing in their power.

Sleep will not come. My body is hot with fever one moment then shivering the next, and I wonder if the night will ever end. A new thought occupies my mind, one of bitterness and regret. The chances are that I'll never know what it is to have a woman, let alone marry one. I want to live, not drown out here in an unknown grave. I begin to question the faith in which I have been steeped. Somewhere along the line something has gone terribly wrong.

Adrift

THE LEADEN overcast sky gradually lightens in the east. Shifting fog banks appear and disappear, drifting and fragmenting as the lifeboat rides the roughening waves. All eyes anxiously search the horizon as we emerge from each trough to momentarily ride the crests. There is nothing, just a desolate grey empty sea and patches of fog. The ship has gone into the depths unseen. No other lifeboats are in sight. The weather and varying ocean currents have scattered or sunk them during the night. We are alone, floating at the mercy of the elements in the black pit.

Despair floods through me as I stare at the grey heaving mass of the ocean. Driven by the wind, sheets of rain squalls sweep across its surface, reducing the visibility to a few kilometres. The galley boy is with his friends; they came for him a moment ago. He lies a few feet away, wrapped in a piece of tarpaulin and a sweater which one of the crew gave him. I don't remember him going. He never said a word, or if he did, I didn't hear him, not with this fever in me. Everything seems dreamlike and now there is no ship, no bloody ship!

The sickness is coming again and I lie hunched up on the seat decking in the bow to relieve the retching. There is insufficient space to lie down properly and my legs dangle uncomfortably in the well of the boat. The wooden planks are wet with the driving spray, but I don't care; I lie on them in a helpless heap in my own world, waiting for the feeling of seasickness to abate.

The background of men's voices suddenly stills and then I hear the captain's voice. I raise my head and shoulders, then leaning back on one elbow, look towards the stern. He stands there to one side of his officers, his hands dangling at his side. A short rotund man with thinning dark hair touched with grey, he's in his late forties, early fifties, but now he looks older as he speaks. "Well," he says, his north country accent dull and flat, "after all these years, the old bitch has gone." He shakes his head, and continues, resignation and bitterness in his voice. "We'd better start back on our tracks. We've a long way to go." He half turns to face the coxswain. "We'll steer due east, Mr. Coxswain, east to Ireland. 'Tis roughly a thousand miles, give or take a few." He clears his throat, his glance taking in our faces as we watch him. There is a tone of finality in the voice. "Aye, lads," he says. "It's going to be hard, but there's nothing else to be done."

A brief silence follows as his words sink in. They confirm what we all know in our hearts is inevitable. Little groups of men, their voices raised or lowered in the intimacy of conversations with friends, argue heatedly as the realization of our true plight becomes apparent.

"Captain!" The Tyneside accent cuts through the background noise. The captain's head jerks up. A middle-aged stockily built man gets shakily to his feet in the well of the boat. An unnatural hush descends on the lifeboat as it wallows in the waves, swinging on the sea anchor. "The rescue ship?" His voice is hesitant, trembling. "Surely it'll come back. They just won't leave us here to die?"

We listen intently for the captain's reply. His short, dumpy figure half leans against the stern planking. He sways with the movement of the boat, his face puckered in thought. "No, I'm afraid not. If they were coming, they'd have been here last night. It's about twelve hours since we were hit, and we were fifteen miles behind the convoy. No, they won't come back, it's policy." His voice reaches us in the bow above the slapping sound of breaking waves against the sides of the lifeboat.

He pauses briefly and continues. "The convoy's a good hundred miles to the west of us by now, and the escort is too busy with the U-boats. I'm afraid they have left us to fend for ourselves."

The captain's words are expected, but the impact still strikes home. It is bitter to take; we have been written off. A deep hurt feeling fills me. The same feeling is reflected in the faces around me. I am not alone. I understand, but my emotions do not. The hurt is too deep and my mind rages irrationally at his words. *Fucking navy bastards, they don't give a shit.* I know in my heart this is not true, that they have no option, but it is a reaction to the old rivalry that exists between the Merchant Navy and the Royal Navy. *The goddamned snotty bastards in the RN, especially their officers, look down on us as a disorganized rabble.*

There is a moment's silence as the captain glances down, gathers his thoughts and continues. "Right now we've got to check the lifeboat stores for food and water. Then we'll make an inventory and agree on a rationing system."

He turns to the small group of officers and men near him in the stern of the boat, and details crew members to carry out these tasks.

I'm out of the way up in the bow, and the movement of men searching the lockers below the seating around the gunwales does not disturb me. A wave of sickness surges through me and I retch green bile. I've had no food to eat since dinner on Christmas Day, and it's now the morning of the 28th. My stomach is empty; there is no food to bring up but the spasms of retching continue. I struggle to the gunwale to be sick over the side, but it is no use, I cannot vomit. I feel the salty spray freezing my face and the spasms cease.

My throat is dry, caked with vomit and it has a vile taste. I roll onto my back, trying to catch the rain which beats on my face with my open mouth. My tongue feels the gap in my front teeth as I do so, and it brings back memories of my sister. I half smile, tears in my eyes as I see her face and, in my delirium, talk to her. "It doesn't matter, Eileen, it doesn't matter."

It was not her fault; we kids were playing. She was under the kitchen table and I was running around it, when she caught my leg. As I fell, I hit the fire guard in front of the Dover stove. My front upper tooth broke off at the gum line. I remember her voice as she said tearfully, "I'm sorry, Mum, I didn't do it on purpose."

"Yes, I know you didn't, love," Mother said to her, as she picked me up and soothed me. The tooth was removed completely last year and I had a small dental plate with one tooth fitted. It was on the washstand early last night, when I went to bed. Now it's gone, with the rest of my belongings.

The rain is cold on my face. I burrow into my overcoat to get warm, and I lick my lips, moistening my tongue and trying to swallow. It brings a measure of relief to my throat.

"Sir, Chippy's swallowed oil!" The urgency in the voice startles me and I sit up; it's the carpenter's mate. He is holding Chippy, the ship's carpenter, in his arms.

He looks at the captain, his eyes anxious. "I made 'im throw up last night, but he's real sick."

Chippy's face is haggard and drawn; oil streaks his hair and clothing as he lies huddled in his mate's arms. The captain is with him now and speaks urgently to the carpenter. "Chippy, can you talk, man?"

I cannot hear the reply, but the captain looks up. "Bring a bucket, he can hardly speak. We've got to make him vomit." He forces his fingers down Chippy's open mouth then thrusts the man's head over the bucket. Chippy gasps and vomits oily slime into it. His head flops back as his mate puts his arms round the frail body.

The captain shakes his head, "I hope none of that oil's got into his lungs. If it has, then he's in real trouble." He looks at the carpenter's mate. "There's not much we can do right now except keep him as comfortable as possible. How did this happen?"

"I was right behind 'im, sir," a crewman says. "He must 'ave missed his footing on the ship's ladder and fell into the sea. We 'ad to pull 'im aboard the boat. He was 'alf drowned and covered in oil, he was, sir."

The captain nods, "Try and keep him warm. If we can find a piece of tarpaulin that'll help keep off the rain and spray."

The men around me stir and begin talking to each other as the captain resumes his seat and speaks to the deck officers. My eyes, dazed and uncomprehending, follow the long waves, as the boat rides up the sloping walls of water to crest the peaks and then wallow in the valleys.

The hum of voices falls silent again as the captain rises from his seat and faces us. His voice carries to me in the bow of the boat. "Men," he says, "until we can rig up a sail, we'll have to draw up a shift roster to share the rowing duties. We'll set an easterly course on the compass for Ireland. Later today, we should be able to hoist a sail, at least a temporary one."

"What about the lifeboat that we're towing, Captain?" one of the crew queries.

The captain glances back, first at the lifeboat barely afloat as it strains and jerks on the tow line, and then at the threatening clouds. The wind has freshened and the seas are getting higher; a storm could blow up suddenly in these conditions. His eyes narrow as he weighs the alternatives, and then, the decision made, he faces the crew and replies, "Yes, we've no option but to cut it adrift. It could easily smash into us in these seas. If it holes us we'll be in a godawful position. We'll do it right away." He turns and speaks quietly to the coxswain. The waterlogged boat is cut adrift.

The remains of the lifeboat float off, floundering without direction in the heavy seas. I lose sight of it as it disappears in a trough of a following wave and vanishes in the vastness of the sea.

"How long do you think it's going to take to reach land, sir?" asks another member of the crew.

"Probably twenty-two days at least, but it depends on the weather and currents."

With my mind and feelings still numb from the torpedoing and feverish with flu, I can scarcely take in all that is happening. The situation is unreal, like a bad dream. The captain's voice seems to come from a long way off as he orders the crew to haul in the sea anchor. With the sea anchor stowed in its locker and a fresh crew at the oars, we are on our way to Ireland. The long, thousand-mile journey has begun.

An engineer examining the lifeboat engine exclaims angrily. "What the fucking hell, there's no bloody petrol in the tank." He turns to a deck officer, and demands, "Are there any jerry cans aboard, or has that been forgotten, too?" A search of the lifeboat compartments reveals none. In frustration, the engineer kicks the engine. "Then we might as well throw this fucking thing overboard; that'll at least give us more room. It's useless without petrol."

The captain agrees but postpones the dismantling until the next day. An even more important omission has occurred. We sit in stunned silence as the captain announces the discovery. His face is angry. Concern puts an edge to his voice, for it is a difficult admission to make. "The bloody storage lockers are practically empty." His voice falters, then he continues, "There are no provisions aboard, and no water, just a few mouldy biscuits. Water isn't a problem. We can catch rain water in a tarpaulin, but food? That's another matter." He leaves the rest unsaid.

We sit in stunned silence for a moment, then, as the implication of what the captain has just said sinks in, an angry muttering grows. A voice grates, "Christ Almighty, Cap'n, what'll happen next? Who the hell's to blame for that?"

The captain's face changes colour, visibly reddening. "It doesn't matter who is responsible right now," he says. "All that matters is getting out of this bloody mess; we'll sort out the blame later, if we make it. Understand?" His voice is angry, and the crew quiet down. He's the captain and they're used to taking orders without question.

"Aye, yer hear that?" a crewman near me mutters. "With twenty-two days to go, and no food? It's as good as a death sentence. Somebody should be hung up by his balls."

"Perhaps we should go fishing without any bait!" a voice adds sarcastically. "Pity we'd not checked the other bloody boat before she was cut adrift."

His friend replies, "Aahhh, what's the point, Jim. That boat was waterlogged. It's more than likely any food was destroyed by sea water."

His patience exhausted, another crewman turns on them and snaps angrily, "For Christ's sake, you two, shut up. If you're so bloody smart why didn't you suggest it to the Old Man? Bitching won't bring it back! The fucking boat's gone, and that's it. The best thing we can hope for is to be picked up by a passing ship." His voice drops, reflecting his despair. "Pretty slim chance of that happening," he says. The small group subsides into silence, each man alone with his thoughts.

There are no further recriminations, no wild accusations. What's the point of them? The lack of foresight in this most elemental precaution of checking and provisioning lifeboats is something I can't understand. Our plight looks hopeless. Depression settles over the boat like a cloud.

Bitterness and despair fill my soul. My eyes dully take in the rowing crew, seated side by side, one pair behind the other, six in all. The rowlocks creak with the strain as they dip their oars in unison and haul back, forcing the boat forward a few more yards with each thrust.

Dick, the chief radio officer, calls to John and me. He has pulled the emergency transmitter from the well of the boat, and placed it on the seat beside him. "The sea water's drained out," he says. "Let's try it. With luck the salt water hasn't corroded the components."

We unscrew the big knurled knobs clamping the accessory panel onto the transmitter case, and tie the antenna to a raised oar. I crank the generator handle while John rotates the switches. The handle moves freely but the instrument remains dead. There is no response.

"Maybe the contacts are corroded, Chief," John mutters.

The chief shakes his head. "Well," he sighs, "We'll have to take it apart and dry it with a cloth later. Leave it for now. Perhaps it's not completely dry inside. I'll give it until tomorrow and then we can try again. It should be dry by then." He sighs. "The bloody thing was supposed to be waterproof."

John joins the chief in the stern and I make my way to the bow. The fever and motion of the boat are bad now and I need to lie down, I don't care where, because it relieves the sickness.

At least we have pieces of tarpaulin aboard. The crew have rigged up some on the weather side of the lifeboat. They give us a little protection from the salt spray, but there's no escaping the rain squalls which soak our clothing. Those squalls come like a moving mist. I watch them as I lie shivering on the wet planking. My feet are numb with the cold, and I try rubbing them to bring back the feeling.

Lying there in the boat, I lose all sense of time. I have no watch. It is our first day. The sky is overcast and threatening. Tonight will be our second night. My thoughts play tricks. Time is not important. I am alone in this sprawling group of men, and speak to no one. Bitterness eats into my soul as I alternate between shivering one moment and sweating the next. *There are no friendly hands to shake my shoulder and ask me how I feel, not even Dick or John. Maybe they came when I was too far gone with the fever; if so I don't remember them, just the emptiness and loneliness.*

A commotion penetrates my subconscious mind, and I sit up with a start. Low moans and stifled screams are coming from one of the men in the stern of the boat. He's doubled up, his hands clutching his groin and is in obvious pain.

Concern spreads rapidly from man to man. I ask the crewman next to me, "What's happening?"

"It's one of the engineers. He can't piss. If his bladder bursts, he could die."

Amongst the small group of officers crowding round the engineer is the captain, who is working over him, massaging his stomach, trying to get him to relax. I cannot hear the conversations but their actions are self-evident. For a few minutes, it is critical. Then, the engineer urinates into a bucket; the danger is over.

The captain stands up and looks round. "Now, listen," he says. "We're passing a bucket around; if you want to piss or shit, do it in the bucket. We don't want any more of these problems. Shock and nervous reaction can, for some reason, close up the urinary passage. That's what happened to the engineer. Fortunately, in his case, he managed to piss. He's lucky, so take care. If your bloody bladder bursts, that's you finished."

The bucket passes from man to man and then the contents are thrown over the leeward side of the boat, primitive but effective. There is no toilet paper. We are down to the basics; survival is all that matters.

I've had nothing to eat or drink for a couple of days, and pass the bucket on without using it. The rain on my face cools my fever and I can taste the water on my lips. I don't seem to need food; the sickness and fever are running their course.

Perhaps it is the delirium set up by the fever as I sweat then freeze, which drives me and creates illusions. I feel alienated and withdraw into myself. Despair deepens and fills my mind as I lie in the boat, a prey to my nightmares.

A noise disturbs me and I struggle to a sitting position, leaning against the gunwale. It must be midday, for several hours have passed since dawn. A stoker is swaying on his feet, and he's shouting at the captain. The sweat rag is still around his neck. Panic distorts his features and he is becoming irrational as he yells, "You're sailing in bloody circles, Captain, you don't know what the fucking hell you're doing. I've a wife and kids back in Shields, man, and I wanna get home."

The Geordie accent, harsh and jarring, causes a momentary diversion from the monotony of the faceless, barren sea. Heated words follow but soon subside as his friends pull him down. "Don't be daft, man, sit down." He settles down again in his seat, still mumbling to himself, but at least he has calmed down.

I sit, partly crouched up, clear of the oarlocks and jammed together with other crew members in the forward quarter of the boat. The stoker's words still ring in my ears, and I understand his fears. I, too, want my home. The old dilapidated farmhouse swims before my eyes, replacing the discomforts, forms and faces around me. The sea changes into the fields and forest that surround it. No electricity or gas illuminates our home; kerosene lamps and candles provide light. There is no running water; butts catch the rain which runs down pipes from the guttering on the eaves of the roof.

I see my father, a wooden yoke across his shoulders. Hanging down either side of him are two three-gallon cans which are attached to the yoke by small lengths of chain. He is tramping the half mile across the fields to the spring, the source of our drinking water. Then I see his dead body lying on Grandma's couch. Tears fill my eyes as I realize that maybe I will be with him soon.

Visions of home replace the monotony of the sea, the crash of waves and the wildly heaving boat as we row painfully back along the path we have come. Visibility is poor. We sit a foot or so above sea level. In the troughs I see only the close sloping walls of water. On the crests, I see the mists and rain squalls advancing across the sea.

I close my eyes. I'm tired and my mind wanders again. It's Eileen, my older sister. The face is vague then clears like the mist. I feel close to her. She's married to an army gunner and lives at home while he is overseas. She has a baby daughter now. Homesickness sweeps over me. I long to feel the familiar forest path beneath my feet, to hear and feel the crackle of leaves as I walk under the great shadows of the beech trees.

The memory of Mother, the night after my father's funeral, flickers and comes back. I remember that night; it's imprinted on my memory because it changed my life. Mother was forty-two when dad died, but her face, lined and worn, looked old in the light of the oil lamp hanging on a beam in the living room. I sensed the underlying urgency in her voice when she spoke. "Aye, love," she said to me, "I've got to get out of the house." She shook her head, talking quietly to herself. "Oh, Jim, Jim, why'd you leave me?" Her despair was evident in her voice, and her red swollen eyes. She glanced up. "Let's you and I go for a walk. The boys are in bed and Eileen's looking after things."

Deep inside me, I sensed that somehow my life was about to change, and I followed her outside into the summer night. There was no moon, but the well-worn country path stood out clearly in the light of the stars. Mother did not speak, but held my arm as we passed the tall elm trees by the little stream and crossed a wooden bridge. It was not until I had helped her over the stile in the fence separating the golf course from the field that she spoke, and then it was not on the subject which I knew was on her mind. "Jim's gone now, love," she said. "I just don't know what to do without your dad. Remember him trying to teach you golf on this field?"

I nodded my head, tears in my eyes at the memory, and felt her hand tighten on my arm. Then she continued. "Eileen's married and Wally's in the army so she's all right, but it's the two boys I'm thinking about. They're still at school and now with your father dead, I just don't know what to do!"

"Don't worry, Mum," I said as I hugged her. "Once I'm at sea, I'll be getting a good salary. We'll work something out."

The lurching of the boat brings me out of my reverie, and cold sea spray carried by the wind beats against my face as the lifeboat crests a wave.

I settle back, huddling up on the wet planking, my thoughts incoherent as I mutter to myself. "No, no, Dad, I've got to stay alive. Mum needs me. Please, Dad, help me." The fever takes over and my mind blanks for a moment. *Yeah*, I think as the spasm dies down. *Mum and the boys should be all right now. I've set up a joint bank account and given her full signing rights. My salary is paid into it each month.*

The afternoon wears on. The threatening storm has held off and a ragged sail is hoisted on a makeshift mast. It helps a little, but we decide that perhaps tomorrow we can make a better sail. Meanwhile a new shift of rowers replaces the old and they bend to the seemingly eternal task of rowing through the never-ending seas.

The Ghost Ship

THE VOICE, loud and clear, jolts me out of the stupor into which I have fallen. I look at the crewman. He's standing on a seat beside the temporary mast. His legs are apart, and he sways as he clutches the mast with his left hand. His face is alive with excitement as he yells. "There's a ship! Look, a ship. Over there." He gestures again with his right hand, pointing to the starboard for'ard quarter.

Amidst growing excitement, we all look in the direction of his pointing arm. We see nothing but the floating sea mist. The lookout's face is animated, his eyes excited as he insists, "I saw it, I tell you, I saw it. It was there one moment, then disappeared in the fog bank again."

Now all eyes are scanning the limited horizon intently. Mist and fog are beginning to gather as a prelude to the frightening darkness of another night in an open boat. It is mid-afternoon, an hour or so before the short daylight hours of the North Atlantic will draw to a close.

The lifeboat wallows in a trough; then, as we ride the crest, we see the ship. Its stern is disappearing into the fog bank. Shouts erupt from the crew. "There she is." Frantically waving to attract attention, we row towards it, but another fog bank hides it from our view. The ghost ship, tantalizingly close, disappears as we reach out for her. Hoping against hope that we have been seen, we continue rowing towards its last position.

To our great relief, we see the ship again outlined through the mist. It has turned and now we see the vessel clearly. It's a tanker. The ship's crew have seen the waving hands and shirts in the lifeboat and changed course towards us. Word spreads rapidly as the vessel nears us. We all recognize her. It's the ship we thought had sunk. Unbelievably the *Scottish Heather* is still afloat, limping along with a great gash in her side.

We are lucky. If it had been dark, we would not have seen the ship, because she carries no lights. We would have slipped by without knowing of her existence, and sailed into almost certain death.

The *Scottish Heather* lies dead in the sea, hove to and ready to pick us up. We row around to the leeward side, and, sheltered from the worst of the weather by the bulk of the ship, approach the Jacob's ladder slung down the ship's side. Ropes thrown from the deck are quickly lashed to the swaying, pitching boat, securing her firmly. Picking their moment, the crew step or jump, one at a time, from the boat onto the Jacob's ladder, then climb up to the safety of the deck.

Now it is my turn. When a wave lifts the boat up I catch the rope sides and jump. My feet find the bottom wooden rung and slowly I climb the rope ladder to the deck. My hand grips the edge of the deck bulwark. I feel a hand on my shoulder and look up into the face of the second mate as he helps me over the bulwark to the deck.

His face expresses concern as I stand shakily on the deck, my feet on the solid, unyielding deck. "Thank Christ, I'm back aboard, Mr. Crook." My head sags wearily, then I look up, my eyes troubled. "I'm sorry I didn't come aft to your boat, but everything was in confusion. At the time, it didn't seem to matter." I pause a moment, then go on. "It was the catwalk. The smoke and noise scared me. I asked my chief whether I should go. He said it didn't matter and that I could help John with the transmitter. Well, I guess you know the rest."

"Yes, I know, Sparks." His face breaks into a smile. "At least you know now what it's like to be in the boats. That's something you won't forget. If you'd stayed with me you would have saved yourself a lot of trouble. We had bacon and eggs for breakfast," he grins and adds. "Well, powdered eggs." The second mate pauses a moment, lays his hand on my shoulder, and continues. "Well, son, you'd better go and have a shower and get some dry clothes on; you look all in." He turns back to the task of organizing the crew to hoist the lifeboat aboard.

My legs are unsteady when I walk across the deck to the companion ladder. The metal handrail is cold to my touch as I grip it with my bare hand. I'm on familiar ground now. An upsurge of energy quickens my pace, and I mount the sloping steps and cross the boat deck to my cabin.

The warmth tingles my skin and I rub my eyes as I enter the passageway. The heavy steel outer door closes behind me, and shuts out the frigid air that has been my constant companion. It is good to be home, insecure as it is.

I pause a moment at the doorway of my cabin, and look at the desolation within. The ship's main generator is running and the lights are on. The door is still lying where I had pushed it. *Is it really only twenty-four hours since the torpedo hit?* It seems an eternity. My washstand and mirror are shattered; glass is everywhere. The drawers beneath the bunk have spilled open and the cupboard door hangs ajar, partly jammed by clothing which has fallen off the hangers.

Aahhh, yes, I say to myself, as I look at the washstand. My tongue flicks instinctively over my teeth, feeling the gap in my upper jaw. *Yes, that's right, I remember putting my plate on the washstand before I went to sleep.*

I half smile to myself; I must be feeling better. The loss of the plate didn't worry me in the boats, but now my vanity is piqued. I scrabble on my hands and knees among the debris searching for the small plate with its one solitary tooth. It is gone. I can't find it in the debris strewn around in the cabin. Perhaps it will turn up later when I clean up.

Sitting on the floor amid the ruins, I lean back against the base of the bunk. My eyes close and some of the tension drains from my body, partially releasing the accumulated stress of the last twenty-four hours. I fall asleep on the floor. *I'm going home, I'm going home*. Cramped and aching I wake with a start. How long I slept, I have no idea; probably a few minutes or maybe an hour. Stiff, I stagger to my feet and stretch, trying to get my muscles working again. I'm still wearing my wet clothes. My pyjamas are sodden and my feet are sore and chilled. I strip off my pyjamas, dropping them on the floor together with my salt-stained bridgecoat, then head for the showers. The hot water cascades over my body, warming my skin and sweeping away the dirt and grime of the boats. Clean dry clothes and a meal boost my spirits and I feel the pulse of the engine through my feet. The ship is heading east; it is heading home.

That evening the hubbub of talk in the dining room is animated and excited. I listen attentively, and, bit by bit, piece together the wanderings of the *Scottish Heather* on the night of the torpedoing.

The captain, mate and second mate are meeting in the chartroom on the bridge. Of the deck officers, only the third mate is in the dining room, and he has an eager audience. "Yeah," he says. "We were all ready to shove off. I was in number four boat tied up to the stern on the port side. Number three was also in the water, ready to go. The second took a final look around and felt the ship's tilt was stabilizing. He called down to us, asking for volunteers. That put the cat amongst the pigeons! Finally, ten of us joined him and the rest cast off in number three boat."

"Aye," an engineer says. "Young Tommy, the greaser, and I spent all bloody night in the engine room getting up steam for the morning. 'Twas no bloody fun. There's only an inch or so of steel between us and the fucking ocean. We'd a' been goners if a torpedo had hit the ship in the arse, man."

The third nods. "You're right there, Jim, I'm glad you volunteered. I'd hated to have been in that bloody place."

A young junior officer asks hesitantly, "I'd heard that the *Chilliwack* came back later. Is that so?"

"Yes. The second mate and I were on the bridge. She circled us and called us on her signalling lamp. She was not sure whether there was anyone aboard. The second answered. He told them we were holed badly in several tanks, but thought we could make the U.K. He said that we'd search tomorrow morning for survivors."

"Did they get the U-boat?"

"No, the *Chilliwack* lost contact with it. They wished us luck and said they would make a report to the escort commander. That was the last we saw of them."

The third jokes, "Well, at least we were warm aboard, better than you bastards. It must have been bloody cold and wet out there in this weather."

"Humph," snorts another engineer. "You'd have been a bloody sight warmer if a torpedo had hit the bloody ship. It's lucky for us that you stayed aboard. We were hoping to see the ship in the morning and reboard her. But she'd drifted out of sight. We thought she'd gone for sure."

"Well," continues the third, "you've got the second mate to thank. He organized the search pattern and everything. With the prevailing wind and weather from the west, we reckoned you'd set a course for the U.K. The second was dead right and we managed to pick you all up by dusk."

"Aye," says one of the engineers, "It's lucky you found us when you did. We'd a' missed you in the dark, that's for sure, man." He pauses and adds, "That and the bloody fog banks."

The third mate laughs and says, "Yeah, I guess so. Visibility, even in daylight, is only two or three miles at the most. At night, unless you had the life jacket lights on, it's pretty hopeless."

A momentary silence follows as we digest his words. He breaks the silence with a query. "What happened to the midships boat, number two, wasn't it?"

"Shrapnel," Dick, our chief, answers, "bloody shrapnel. It blew several holes in the bottom of the boat and she sank up to the gunwales. We had to pull her crew aboard our lifeboat; they were very nearly drowned."

I look down at my plate, half-filled with uneaten food, and silently agree. *We were bloody lucky.*

Ours was the last boat to be rescued. With all our crew accounted for, the ship is free to proceed to the nearest home port. The captain takes over the ship from the second mate and we lay a course for the west coast of Scotland and the Clyde.

Through Hostile Seas

LISTING BADLY to starboard, the *Scottish Heather* heads east. Painfully, she makes her way through heavy following seas at reduced speed. We're limping along at seven knots now. The third mate told me that during dinner, the first night back aboard. "Aye," he said to my query. "If we go any faster, the plates will tear off around the bloody hole for'ard." He smiled and added, "Provided all goes well, we'll make port in five or six days at this rate."

Exhausted, I sleep beneath warm sheets and blankets, with a soft pillow for my head. My life jacket is close to hand next to my pillow in the bunk with me. I, like everyone else on the ship, have gone to bed fully clothed, except for my shoes, jacket and tie. I've learned the first rule of survival the hard way.

It is light when I awake to the reassuring rhythmic beat of the ship's engines. I feel rested after a night of uninterrupted sleep. The fever, which reached its peak and broke in the boats, has passed. Now, with my spirits restored, I slip out of my bunk and go down to the washrooms on the deck below. I look into the mirror as I brush my teeth and shave. My eyes have a yellowish tinge, and my skin is also turning a subtle shade of yellow. When I go to the toilet, I really become concerned. My urine is dark, dark as strong tea.

The smell of bacon and scrambled eggs is appetizing as I pass the galley door on my way to breakfast. It awakens my taste buds and I hurry to the wardroom. Dick, my chief, is already there and he beckons me over. "Feeling better today, Gord?" he asks.

I laugh as I sit down and order breakfast from the table steward. "You bet, Chief. I slept like a log last night. A good night's sleep has made a world of difference."

I hadn't realized how hungry I was and eat quickly. Dick grins. "At least you've got your appetite back." We talk awhile, then still feeling concerned about it, I mention the discolouration of my skin and urine.

He looks at me closely. "Yes, you're right. I think you must have yellow jaundice. It's probably through a combination of flu and exposure in the lifeboat."

He thinks a moment, then continues, "You'll have to see a doctor about that. A medical team will come aboard when we get to the Clyde; it's normal procedure. I'll make sure you see them."

The radio room is a hive of activity when the chief, John and I descend on it after breakfast. The main transmitter is damaged beyond repair, but the sturdy little communications receiver still functions, and the emergency transmitter is operational. If we are attacked again we may at least have a chance to get off a distress call. It takes us the best part of the morning to clear the debris, get the equipment into working order and straighten our filing systems.

Dick looks around the radio room as the last files are cleaned up. "Aye, lads," he says. "That's it, she's all shipshape. We're back in business." He turns to me. "It's your watch, Gord, but have your lunch first."

Together we go down to the wardroom. Life is back on track; our watch-keeping duties have returned to normal, four hours on, eight off; normal, except that we are now defenceless. We have to cross a thousand miles of hostile sea with no escort to defend us. There is not a man aboard who doesn't realize that our chances of making it safely to port are slim. We will need luck and the weather to hide us as we thread our way through the net of U-boats along our course to the Clyde.

The chief's neat rounded handwriting stares up at me when I sign on the log. He's written in the preamble on the top of a new page: *Watch keeping resumed. Proceeding UK December 29, 1942. For the crew of the Scottish Heather, this is an historic day. At least we are alive and now, thanks to the second mate, stand a chance of surviving.*

I reach for the headphones and tune in to Rugby Radio. The radio traffic is heavy for the convoy but there are no messages for our own coded callsign. I take the cypher groups down for information and decode them. *Yes*, I think as I put the heavy manuals back into the metal case where we keep them, *it's as well it was our lifeboat which sank, not the captain's. Dick took the manuals with him when he boarded the captain's lifeboat. If it had been sunk, we'd have lost them.*

Putting aside the decoded messages, I call the bridge on the telephone and speak to the second mate. He's coming to collect them. I hear the heavy tread of his footsteps on the stairs from the wheelhouse into the passageway, and glance up as he enters the radio cabin.

"Anything interesting, Sparks?" he says.

I nod. "Looks like the convoy's had heavy losses. I've heard faint distress calls on the medium frequencies, but the convoy's moving out of range on 600 metres.

However, I've taken down these messages for the convoy and they don't look good. I think they're getting hammered."

His face is serious. "Yes," he says quietly, "I was afraid of that." He draws in a long breath and sighs. "Anything for us?"

I shake my head. "No, it looks as if they've given us up for lost. You know, written us off as another casualty of war."

A half smile flickers across his face. "Yeah, could be." He looks me in the eyes, and his face wrinkles into a broad grin. "We'll get there, son, don't you worry. We're going to make it."

Sixteen hundred hours, and I'm off watch. The cabin steward is cleaning up my cabin when I come in. "Haven't found your bloody teeth yet, Sparks, but I've only just started," he says with a chuckle. "You should 'ave worn 'em. They're probably over there under all that rubbish, but mind the glass," he adds when I kneel down to sort through the rubble.

He's right. It's there, lodged in a crack between the bulkhead and the cupboard. "Bugger it," I exclaim as I examine the broken remnants of my dental plate: The tooth has been knocked off and the plate badly cracked. It's useless.

The cabin steward laughs. "Tough luck, Sparks," he says. "You'll just have to keep yer bloody mouth shut."

In disgust, I throw the plate in the garbage, and help him tidy up my cabin. It's amazing what a broom and willing hands can do. My cabin is soon made habitable. A little later, the carpenter's assistant arrives to fix my cabin door and do other minor repairs.

Barely twenty-four hours have elapsed since our rescue from the boats and return to the comparative safety of the *Scottish Heather*. The heady euphoria of yesterday has evaporated, and now the relaxation of tensions and stress are taking their toll on my body. The fever has passed, and apart from the residual effects of jaundice with its yellow skin and dark brown urine, it, too, is almost a thing of the past. That is, it was until today.

Ohh, Christ, I think as I lay in my bunk, debating whether to go down to the ship's head. *I've got to go.* I've just come off the midnight-to-four watch and the need to visit the ship's head is urgent.

I remember with fear my last visit to the toilet. I noticed something that had not registered before; the porthole in the ship's head is small, hardly big enough to put your head through, let alone your body. The realization brings back memories of the *Soborg*, my previous ship, and triggers a vision, buried deep in my subconscious, of men burned to death, trapped half in and half out of portholes like this.

During an attack on the *Soborg*, a bomb hit the stern gun platform and ignited the ammunition locker. The crew members in their cabins had died horribly, struggling to get out of the portholes. Trapped, they were burnt to death in the flames. That ship was in dry dock when I joined her; all those portholes were enlarged, but it was too late to save her crew.

A feeling of entrapment comes over me, my heart starts to pound and there is a suffocating feeling in my throat. It feels as if the bulkheads are closing in on me like a steel coffin. In my mind, I can hear the screams of those men as they burnt to death, and fear drives me to the deck.

Now as I lie in my bunk, I am afraid to go to the ship's head, and instead I slip out of my cabin to urinate over the leeward side of the ship. Unhygienic perhaps, but it is a measure of my apprehension and fear of being trapped below decks.

Ashamed, I tell no one of my feelings, but keep them securely bottled up inside me. *Am I alone in these fears?* I wonder if the other men feel like I do? I only go to the ship's head when I absolutely have to, and try to make sure that's in the daytime.

The *Scottish Heather* ploughs steadily through following seas at six or seven knots. A heavy swell causes the ship to pitch and roll. It's now the second day into our return journey. We're two or three hundred miles away from the battle being fought to our west, but there's still a long way to go. It's cold outside and dark.

I'm off watch and the evening meal is over. For once, I linger in the wardroom, and, lulled by the swaying of the ship, half lie on the wardroom settee. A voice calling my name breaks into my thoughts. I open my eyes, and glance up. It's Brad.

I'm pleased to see the fair-headed young cadet, and make room for him on the settee beside me. "Hi, Brad," I reply. "It seems a long time since I saw you at Christmas dinner, but then, a lot of things have happened."

His face breaks into a smile. "Yeah, it's about four or five days." He hesitates, then continues. "I saw you in the lifeboat; you looked all in."

My interest quickens. "Were you in the captain's boat?"

"Yes, you lot were bloody lucky to reach us."

Anger stirs within me. "The bosun turned us away. He wouldn't let us near them." I catch my breath, remembering vividly that awful moment, then continue bitterly. "He said there wasn't room. Bloody liar. Yeah," I pause momentarily, "we were lucky. If the Old Man hadn't let us aboard we'd have been in the sea hanging onto that boat. The bloody thing was almost up to the gunwales in water. Christ, it was cold. God knows how long we could have taken that."

Brad nods. "The sea's freezing. You know, I didn't realize it was you lying in the bow until the morning. Then it was only when I saw your bridgecoat. The second sparks came aft. Why didn't you come too?"

I look up at him, the hurt still lingers in my mind. "I went to bed with a flu right after my watch at 1600 hours. I'd not eaten all day and I felt bloody sick. All I wanted to do was lie down somewhere in the lifeboat. I didn't care where. I hadn't the energy or will to go aft."

Brad nods his head, then says, "It's never like you think it will be. I mean, the panic and mad rush for lifeboats, the confusion." He shakes his head in disbelief as he says, "Not checking lifeboat supplies, no food and water, then no petrol for the engine . . . Christ, it's criminal." He sighs, "Yes, we act irrationally at times. If it wasn't for the second mate, you and I wouldn't be here now."

Brad sits still for a moment, deep in thought, then looks at me as if wondering whether to say something more or not. He makes up his mind, his eyes troubled.

"I'm sorry, Gord." He pauses, groping for words. "A group of us started a lottery on who would die first in the lifeboat. We thought it would be you. I realize now that was insensitive, and it bothers me. At the time there seemed no harm in it."

I am surprised and bewildered. "But if that's what you thought, why didn't someone come over to see how I was? You know, I shared my bridgecoat with the galley boy the first night. He had only his life jacket, a shirt and trousers. We cuddled each other for warmth. In the morning his friends found him and gave him a piece of tarpaulin and a sweater. They looked after him. Why didn't someone come to help me?"

I drop my eyes; my hands are trembling. There is silence for a moment. Then I say bitterly, "It's not you, Brad, but I would have thought my chief or John would have come over to me. That would have helped." I feel hurt, and deep inside I am angry.

Brad hesitates, embarrassed, then says, "Yeah, I know. But at the time I didn't think about it. The lifeboat was packed so tight with men it was hard to move about. Once you're in a group, you kind of stay with them."

"Well," I reply, "it doesn't matter now. We're back on the ship and on our way home. It's like waking from a nightmare, and hopefully it's over."

I feel weary and tired. Getting up, I leave the wardroom, disillusioned and disturbed. I am angry and further alienated. Brad meant no harm and just wanted to talk; maybe it is his way of showing friendship. It's hard to say. But I don't appreciate the gallows humour. My views on men and the society into which I have been plunged are beginning to change, and now I have further cause to mistrust my own gender.

Shrouded in a duffel coat, its hood pulled up over my head, I watch the sea from a sheltering bulwark under the bridge on the deck outside my cabin. The bulwark cuts the force of the wind and the bridge protects me from the driving rain. There is a nervous tension in the air that has not been there before. The old easy mannerisms have gone, replaced by a certain jumpiness, too quick responses and a hard-to-define edginess. I am not alone in my feelings; we all live with the expectation and fear of another attack.

We are within two days' sailing from the Clyde estuary, approaching the northern coast of Ireland, when we hear the distant drone of aircraft engines. It doesn't sound like a *Focke Wolf Kondor*, but we take no chances. The ship's klaxon brings the vessel to full alert for an air attack. Men run to man the machine gun nests on the wings of the bridge and aft of the funnel.

Tension mounts as the roar of the aircraft engines increases. It heads directly for us above the heavy cloud cover. We scan the overcast skies anxiously. The tone of the engine changes suddenly when the plane dives into the clouds. It breaks through the cloud cover at about one thousand feet above sea level, slightly off our port bow. We relax with an audible sigh; it is a Royal Air Force Sunderland flying boat from Coastal Command.

The aircraft flies low over us, circling us twice; we can see the dim outline of the pilot in the cockpit. Having identified us, he flies in low, parallel to the ship, barely skimming the surface. He waves his hand in salute, before flying off to the northeast.

We watch the aircraft as it climbs steeply and disappears into the clouds, heading back to base to report our position, course and damage to the naval authorities. Soon after the sighting, we receive a coded radio message on the Rugby Radio frequency directing us to Rothesay, a small port with naval docking facilities. The naval dockyards and surrounding town are located on the Isle of Bute in the estuary of the Clyde. There, the navy divers can assess the damage to the hull of our vessel.

Our ship's course is along the northern coast of Ireland, but well clear of the coast itself. The tension relaxes when we near the Clyde estuary. Luck and foul weather are on our side. We have slipped through the network of U-boats lying in the Western approaches of the Atlantic and the coastal waters surrounding Britain.

We arrive off the Clyde on January 2, 1943, fourteen days after joining convoy *ONS 154* off Malin Head. The *Scottish Heather* has completed a voyage of a thousand miles from the point of her torpedoing. The pilot launch lies just ahead of our starboard bow. We heave to, and wait for it to come alongside as the crew lowers the Jacob's ladder. The small launch noses expertly into the foot of the ladder. Standing on its bobbing bow, the pilot grasps the rope ladder with accustomed ease and clambers aboard.

Now in the safety of the estuary, the tanker limps into the anchorage in Rothesay Bay. The great gash in the starboard side is mute testimony to the reason for our return. Ships' sirens and whistles sound off in salute, welcoming us home, spreading the news that we are back from the dead.

We pass slowly into the anchorage, weaving between the ships, their decks lined with curious seamen watching and waving in welcome. It is an eerie feeling to finally make port. I have tears in my eyes and a lump in my throat as the anchors rattle down through the ports in the bow. The long journey is over and the prospect of a month's survivor leave stretches ahead.

With the anchoring of our vessel in the bay, we are invaded by naval and dockyard officials. The medical authorities come on board to check on casualties, and, with several other members of the crew, including the carpenter, I am sent ashore to the naval hospital in Rothesay.

The doctor, an elderly Scotsman, confirms my jaundiced condition. "Aye," he says, "you've got jaundice all right, laddie; it's a combination of flu and exposure that's done it." He pauses and then adds, "It would be best if you stayed here in the hospital a few days. You need rest and treatment, you know."

He laughs when he sees my disappointment, and relents. "Oh, I know, son; you want to go home, don't you. I can't blame you, it's been a pretty nasty experience." I feel his hand on my shoulder as he continues. "I'll tell you what we'll do. I'll give you a note for your own doctor to treat you at home if you want. How's that?" Armed with the letter and medication, I thank him, and wander off down the hospital corridors to the waiting military transport.

There is a flurry of activity aboard. A few days earlier, the government had sent telegrams to our families regretting that the ship was lost, and that we were *missing at sea, believed sunk with all hands.* In addition, the German propaganda radio broadcasts had also announced that our ship, among others, was sunk. We are handed telegram forms to complete and send to our families assuring them that we're safe in the U.K. Among the pile of telegrams handed into the purser's office is my own.

The following day, naval divers inspect the hull of our ship to assess the extent of the damage and to make minor repairs. The vessel can't be repaired on the west coast, so we will have to take the ship back to one of the large dockyards on the Tyne.

Late that night, the remnants of our crew return from shore leave, escorted by the military police. Bitter at being left by the navy to die in the lifeboats, a small group have been in a fight with naval men in one of the Rothesay bars. Emotions mixed with alcohol got the better of reason. The military police had to break up the fight, and our crew return to the ship carrying battle scars to mark their frustrations.

Early the following morning, we join a slow coastal convoy and reach the Tyne without incident three days later. The *Scottish Heather* ties up alongside Smith's Dock, North Shields, the same dock that we left on December 13.

It is January 11, 1943. After signing off, I walk across the decks of the *Scottish Heather* followed by the curious stares of the dockworkers who view our badly damaged tanker. I feel the sting of hidden tears as I avoid their glances. I am afraid to speak for fear of crying. I hurry on, my gaze fixed in the distance. It is the end of the journey. Tired in both mind and body, I board the London train.

Survivors' Leave

THE STATION is quiet when I open the carriage door; only the hissing of steam from the train engine breaks the silence. It's one in the morning. Now, as I step down onto the dark and deserted platform, a dimly lit sign on the grimy station wall indicates it is Chingford, the end of the line from Liverpool Street Station in central London. The blackout is in force, but at the moment the night skies over London are free of enemy bombers.

The ticket collector, a dim shadow behind a small circle of light at the exit, grunts, "'Night, sir." The street is deserted and unlit. I pause, then walk quickly down the street. There is a chill in the air but it's not raining. My eyes are used to the dark now and I can make out the water trough at the Y-junction of the road leading to Sewardstonebury. This relic of bygone days was once used for watering horses. As I reach it, the warbling note of the air raid sirens gives the alarm.

I've a long walk before me, because there are no buses to my home. Home is a small farmhouse tucked away in a fold on the edge of the forest, and was once the gate lodge to one of England's large country estates. The house is buried amongst the trees, half a mile from the road. Ever since I can remember, I walked the three miles to and from school in Chingford, a suburb of northeast London. The station is a mile shorter.

Glancing at the sky towards the east, I hesitate. There are no shell bursts in the sky to warn of approaching aircraft, so I continue walking up the road across the golf course to the forest beyond. I'm halfway to the forest when the first anti-aircraft shells burst overhead. It is several minutes before I hear a sound like the humming of angry bees, the whine of shrapnel from the anti-aircraft fire falling back to earth. I glance up in alarm and run for the shelter of the trees.

My grip tightens on my suitcases as I run. Sparks fly off the hard paved surface of the road when a piece of shrapnel strikes not ten paces from me. Swerving off the road, I reach the trees and shelter in their familiar bulk. I'm breathing easier now as I press against the trunk of a tree. The forest is alive with the sound of falling shrapnel; it crashes through the branches and bites into the wood. The flurry of metal, like a hailstorm, passes on and silence returns to the forest. The shell bursts recede towards the southwest and inner London.

I walk along the paths I used to roam in my younger days. I love this forest; I've lived here, built dens, and listened to its sounds, in summer and winter. My family are woodsmen; to us, the forest is home. There is no moon tonight, but the sky is clear; I pick my way by starlight. The ground is frosty and crunches beneath my feet. I hear the tinkle of water running in a stream and know I'm near our house.

The "all clear" sounds when I'm on the wooden bridge my father built over the stream. There are no handrails on the bridge, and I tread carefully across the thick planks which he nailed in place. I emerge from the forest into the glade. The dark silhouette of the house stands out framed by the tall trees at the end of the clearing. A dog barks as I open the gate. It's old Satan, our black Labrador. He quiets down when he hears my voice.

"Hush, Satan," I whisper to him. I hug his furry coat as he tries to lick me, then I knock on the door.

My mother, roused from her bed by the dog's barking and my knocking, opens the door a few inches. Keeping it on the safety chain, she peers out into the darkness at my shadowy form.

"It's me, Mother," I cry, "it's Gordon."

The chain clinks as she opens the door, saying, "You're home! Gordon, you're home." Half laughing and half crying with emotion, I step inside the old farmhouse. I am back with my family.

Mother's voice is broken, hardly coherent as she hugs me. "We . . . we thought you were dead. The government sent us a telegram. It said . . . it said that your ship had been lost with all hands, and you were presumed dead." She holds me away from her as she looks at me, smiling through her tears. "Then we got your telegram saying you were safe."

Mother laughs with relief. "I knew something had happened to you, I felt it, but then I didn't believe you were dead. You were born in a cowl. 'Tis an old sailor's superstition, love, that you can't be drowned."

She's busy now stoking up the fire in the Dover stove to make the inevitable cup of tea. My two younger brothers, Peter and Douglas, have been awakened by the commotion and after the initial excitement make themselves small in the shadows thrown by the kerosene lamp on the dining room table. I grin at them, knowing full well that they're hoping Mother will not notice them and send them back to bed. It's a vain hope. Mother knows they are there. "Now you two boys," she says, "you've seen your brother. It's late and you've school in the morning."

"But, Mum," Peter begins to protest.

"No buts! Gordon will be here tomorrow. Now off to bed with you, or else."

Grumbling, the boys disappear, their passage marked by the thud of their feet on the stairs. Mother listens to the creaking of the stairs and, as the sounds die down, she smiles, "That's sorted them out."

We talk long into the night. She listens intently as I recount my experiences aboard the tanker. "Aye," she sighs, "we heard about it first on the German radio broadcasts, you know, Lord Haw Haw. He even named your ship, and said you'd been sunk along with a long list of other ships. If what he says is true, that convoy must have really got hit hard."

I nod, "Yes, it was. Four ships were sunk on the first night of the attack; then we were the first to be torpedoed on the second night. God knows what happened to the convoy after that. There were so many U-boats around us, Mum! We were left behind and took to the lifeboats. If it wasn't for the second mate, we would never have got back." I fall silent and then ask after Eileen.

"Wally's on leave and she's with him." Mother pauses a moment, then goes on. "They've gone to visit relatives. You just missed them. He came home on leave for Christmas and he's being posted overseas." A troubled frown creases her forehead, as she says, "These days one never knows where any of you are going. It could be anywhere from the Mediterranean to Burma." She looks up and smiles, "I'll put you in their room; the bed's made up. Eileen and the baby can sleep in my bedroom when she comes back. T'will be no problem, Gordon. Wally's got to report back to the army. He'll not be staying."

I look at my watch. "Do you know the time, Mum? It'll soon be four o'clock; we'd better get to bed." I get up from the armchair and stretch, then almost as an afterthought, I add, "By the way, have you heard from John Cummings? Did he get that job aboard the Belgian passenger liner?" Mother doesn't answer at once, so I say, "You remember; the Belgian shipping company sent me a telegram, offering me a seagoing job on my last leave. I gave him the telegram and wrote to the company forwarding his name in my place. I'd planned on visiting his family some time this week."

Mother hesitates, and looks up sharply, her face troubled. "Yes," she says slowly, "you won't have heard. His mother, Mrs. Cummings, told me. I . . . I'm sorry, Gordon. He's dead!"

Stunned, I feel the blood drain from my face. My legs are unsteady and I collapse into the chair. "Wh . . . what d'you mean, dead? He, he can't be, it was a big liner, a troopship!"

"Yes," Mother replies. "He got the job, but the ship was torpedoed on the way back from the North Africa landings, and sank in the Bay of Biscay." She pauses, her eyes look down at her work-soiled hands, her fingers unconsciously wrapping and unwrapping a handkerchief. Her voice breaks the silence. "I'm sorry, love, but he was last seen in a lifeboat. He was never seen again. Just like you, only you were lucky."

I nod, and cannot answer at first; the words choke inside me. "Oh, Christ, not John! Why, why? I got him that ship!" I get up shakily and stand by the door leading to my bedroom, hesitating, unable to think coherently. "Bloody war," I say brokenly. "Why the hell did I come back?" The hurt and pent-up tension in my body explode and I can no longer hold back the tears. "Christ, he was only a year older than me." The tears ease my inner hurt and I feel anger, real anger. "He knew nothing of life, and now he's dead." My voice falters and I mutter bitterly. "I think you're right, Mother. There is no God, not a loving God. How can it be otherwise with all the killing?" I look up through my tears. "I'm sorry, Mum, I can't think straight right now. I feel worn out."

Mother comes over to me as I stand by the door. I hug her to me, and can feel her body shaking. Tears run down her worn face. She looks prematurely old. "You mustn't blame yourself, Gordon," she says. "John wanted to get to sea so badly. He was just like you. Couldn't wait to get into the war. It's the parents that suffer, too, don't forget. His father and mother are going out of their minds. He was their only child, and now he's gone."

The white bedsheets are warm and soft when I get in between them, and pull the blankets up. But, tired as I am, sleep does not come easily. The memory of John, looking like a wise old owl with his sharp angular features and large horn-rimmed spectacles perched on his nose, comes back. He was eighteen, young and full of life. John and I did our training together at the Radio College in the winter of 1941-42. He lived in Walthamstow, on the High Street. I've known his parents for over a year now.

Once started, the memories flood back, and with them the feeling of guilt, of having signed John's death warrant. *If I hadn't sent in his name, he would still be alive.* It's an irrational and illogical thought, which doesn't ease my feelings. That last leave is so clear in my mind. John's delight at getting a ship, a passenger liner, was evident; he had grinned and jumped at the chance. We spent that leave together exploring the theatres, dance halls, and nightlife of London. That was the last time I'd seen John, and now the sense of loss sinks in deep as my mind goes over the past. My thoughts are jumbled, one tumbling over the other, until finally I fall into a troubled sleep knowing that I'll never see him again.

In those first few weeks, the jaundice runs its course under the watchful eyes of our family doctor. It is not the jaundice that worries me, but the nightmares that torment my sleeping hours. They bite deep into my mind following an endless pattern. How can I forget them? They come back to me again and again as I lie awake in bed. I am drowning; a scream wrenches from my belly and choking sensations rip through my body, only to be replaced by other sensations. Like the eyes of a nightmarish devil, the little rows of red lights appear and disappear in the dark of the night. Surging seas throw the lifeboat up, then slam it down, forcing its gunwale down into the sea. I feel the icy cold sea closing in on my body and shout in terror, half screaming, then wake in a sweat, my heart racing.

My mother comes into the room. "It's all right, Gordon, it's all right," she says. "You're home now. It's only a dream." She sits there until, regaining my composure, I calm down.

"Your mother is a communist, a card-carrying communist." Brother Isidore's voice is harsh and grating. Feeling a flush of shame, I start to protest, but he holds up his hand, palm out, silencing me.

"Let me finish," he says. "This is the reason I didn't give you a sacred heart when you left the college three years ago. Because your mother is a godless woman, it just didn't seem right." He pauses a moment, his eyes glittering behind the small rimless glasses. "But perhaps I should give you the talisman, especially after your recent experience in the Atlantic. You could have lost your life when your ship was torpedoed." Brother Isidore is silent a moment, his eyes downcast, hands fumbling in the folds of his gown. "Four ships sunk before yours! How many more?" He mutters the words to himself, leaving the sentence unfinished, then glances up. "Here, take the talisman. The sacred heart may protect you."

He holds out the small blue box. It fits snugly in the palm of my hand. Bewildered and shocked by his words, I am unable to think clearly. These men in their gowns have always overawed me; they are like gods. In the old religious college they ruled through the fear and wrath of their God. He was not a loving God. Perhaps I should refuse the talisman, but the magnetic hold of the Order is still too strong.

It is early February, and, although the day is cold, the sun is shining fitfully through thin scattered clouds. On impulse, I had decided to visit my old school, and had put on my uniform; I take pride in wearing it because it makes me feel like a man. But Brother Isidore has made me feel like a child. He is the principal of St. Egbert's College, a Catholic college run by the Order of the Sacred Heart. He stands there in the entrance hall as I open the small blue box. The red heart lies against the blue and white lining of the case. The velvety material feels soft to my fingertip. A fine gold thread outlines its edges.

"It is a bleeding heart, the symbol of our Order," Brother Isidore says. "Take care of it; it may save your life."

I nod my head and turn away as he says goodbye. Glancing back, I see him for the last time as I cross the school quadrangle. His bulky body is cloaked in a dark brown, ankle-length gown which almost touches the floor; a cowl hangs down the back of his neck, and a large red cross is sewn down the front of his gown. His heavy jowls partly hide the stiff white clerical collar.

It is dark by the time I get home. Dusk falls early in midwinter and I've been delayed at the College talking to old acquaintances. The back door creaks when I open it and enter the scullery of the farmhouse. Mother's preparing supper and I smile as I hear her cussing out the kids. She's not changed in that respect. "That you, Gordon?" she calls. "Dinner's almost ready."

Later in the evening as I sit alone by the log fire in the living room, which now serves as my bedroom, I turn the talisman over in my hands and ponder Brother Isidore's remarks. I smile as I remember the communist rallies held in our field before the war. *Sure, she's a communist. Why shouldn't she be? Educated too, she's the businesswoman of the family and made a profit from all those rallies.* Mother's tea, cakes, sandwiches, and ice cream sold like hotcakes. She managed the money end of our small camping and catering business. True, it was primarily a weekend business, but it was a profitable one, especially during the summer in those days. Dad worked as a landscape gardener for seven shillings a day and got a bit more on weekends as the local golf pro on the nearby golf course.

The flames shoot higher as I feed the fire with the logs I'd cut and split. Since my childhood, flames have always fascinated me. My father used to sit on this fireside seat where I now sit, poking the logs into flame. I ask myself, *Why, why did he have to die? There's so much I would have liked to ask him.*

My thoughts turn to Brother Jarlath; I can see his face in the flames. I liked him; he was kind while Brother Finbar was brutal. But then, what price was that kindness? Brother Jarlath liked little boys, and, unfortunately, I liked him. I'll never forget his face, every bloody wrinkle on it, his large nose, the thinning auburn hair and blue grey eyes.

My eyes cloud over as I think back on my school days. It was spring, a warm spring day towards the end of my first year at St. Egbert's, sometime in 1936. I remember the sun shining through the large classroom windows. The school bell had rung and it was time to go home. Forever a dawdler, I was the last to leave. The other boys in their brief shorts and knee-length socks, their school blazers and caps hurriedly thrown on, had left, closing the door behind them. Only Brother Jarlath and I remained while I loaded my satchel with my homework.

He waited for me patiently, an amused smile on his face. Then he put his arm around my shoulder as we walked to the door. He often did that to the boys. But this time, without warning, I felt one hand slide up the wide leg of my shorts as he gripped me with the other. I saw his face as he bent down. His eyes were strange; they seemed to glisten with an inner glow, not the eyes that I'd come to trust. Then, I panicked as his hand felt beneath my shorts, his fingers feeling between the cheeks, touching my anus.

"Don't be afraid," he whispered, as I tried desperately to escape his grasp.

I'd never felt anything like it before and terror lent strength to my arms as I struggled. How I broke away, I don't know, because my mind blanked out and I found myself far from the school, shaking with shock and out of breath from running. My chest was heaving when I reached the forest path which led to my home.

As the forest closed around me, I calmed down and lay on the soft grass in a glade. The late afternoon sun was warm and soothing. *Brother Jarlath*, I shook my head, turning things over in my mind. *Who'd believe me, the poor kid in the school? They'd laugh and I'd feel foolish again.* Most of the others were boarders, rich kids, but Mother paid a reduced fee. She'd made a deal with them, because she felt I'd get a better education at the college. My heart stopped pounding, and I was tired.

I kept it to myself; I didn't tell my mother or my friends. I felt soiled, ashamed. Brother Jarlath was my form teacher and, after that experience, I kept him at a distance.

The voices of my mother and sister drone on in the background as they iron clothes, heating the irons on the old Dover stove. Eileen has come home. I saw Wally just before he left to return to his unit. He's the same old Wally, tall and well-built with a permanent grin on his angular face. The short cut wavy black hair is brushed back from his broad forehead with its thick black eyebrows. He looks fit and smart in his army battle dress. Eileen looks short beside him; she's about five two or three, while Wally's six feet four.

Now as my thoughts return to the present, I poke the logs in the grate and watch the sparks flare up. They seem angry and in harmony with my thoughts as I think back on my childhood because, apart from Brother Isidore and the talisman, I learned many things this morning from old school friends and acquaintances. Brother Jarlath had been found guilty of molesting young boys and sent to another school in the north of England. Brother Finbar had left the Order to join the Royal Air Force as a fighter pilot in 1940; he was shot down and killed over Britain. I remember the savage joy I felt on hearing of his death. The depth of my anger surprised me, but then, he was a brutish man. His brutality taught me to hate, and that hate still lingers in my mind.

Memories of nightmares I've had since my childhood come back and now I try to analyze them, to ponder the mysteries of the past. It's almost an obsession. I search for the truth, an elusive truth that drifts away before I can grasp it. Picking up the talisman, I look at it, plead with it for peace of mind, but that feeling of peace does not come. Perhaps it never will.

Always in the nightmares, there's a vague figure in the darkness of the stairwell, but I can't make out who it is. I feel my arms held so tight that I cannot move them and I'm screaming, "Mummy, Mummy." It is a child's voice, a frightened voice. There's a distant light, a faraway round light, as though I'm in a tunnel, and then, I wake in a sweat.

I shake my head and put the talisman back in its small blue box. Those dreams are so vivid. I don't get them so often now, only occasionally when I'm stressed out and depression sets in.

The voices in the dining room are quiet, and I hear the creaking of the wooden stairs leading up to the bedrooms where my own room used to be. It's probably Eileen. There's no electricity in our house, just candles and kerosene lamps.

I remember as a young child, what a ghostly place it was when the wind whistled around the roof and shook the windows. The house seemed to bend and creak in the wind and the candle's flame wavered, casting moving shadows on the walls.

As I sit beside the fire, brooding over the past, I remember Mother's familiar voice, and the reassuring light of her candle when she opened the door to my room. "What's the matter, Gordon?" she used to say.

"I'm frightened, Mum," I'd reply.

She'd smile and reassure me, and leave the candle with its flickering flame in the candle holder on the bedside table. "There, that will make you feel better. You're frightening yourself; there's nobody under the bed. Now get to sleep; you'll wake your brothers."

"You've changed," Mother said earlier tonight, when I told her about Brother Finbar's death. "There's no room in you for pity anymore. He died a hero."

"Yes," I said bitterly. "He may have, but that does not make him less a brute. I hope he rots in hell, and Brother Jarlath for that matter. Did you hear about him?"

"Aye, there were many rumours at the time." She paused, her face thoughtful, then continued, "He got caught molesting young boys at the school. Aye, I heard that. It caused quite a stir at the time, but the Church hushed it up." Mother shook her head in exasperation. "They dealt with it and he was sent up to another college in the north. That's like putting a wolf in amongst the sheep," she said. "Humph, celibacy! It's not natural; it sends them bloody crazy in the end."

Mother's words linger in my mind as I lie in bed, watching the flickering shadows on the ceiling cast by the dying flames. I wonder if she knows or even suspects how troubled and angry I am. The loss of her husband has changed her, too; not that she was ever a loving and caring person. Dad just about disappeared into the woodwork, the strong silent type. Mother reckoned that he couldn't deal with children; she carried out the discipline and firmly believed that to spare the rod spoilt the child.

My thoughts are too active for sleep and I find myself thinking more about Mother. She and Dad were so unlike each other, but there's no doubt who was the stronger spirit. Born in Easingwold, a small village not far from York, she was well-educated for her day. She was a Dunn from an old Yorkshire family, once wealthy in the heyday of the north country cotton mills. When the cotton industry failed, the family fell on hard times.

Granddad Dunn was the stationmaster of York railway station, quite a prestigious position. I remember him well; he was a short, stocky man with a Hitler moustache. He and Mother were kindred spirits but never seemed to get along when he came to visit. On one occasion, Mother gave him my dad's shotgun, and said, "Take Jim's gun and go and shoot your bloody self, you old bugger. Go on, do it in the forest." Then she added in a lower voice, "You won't mess up my kitchen then." I don't know what the row was about for I was very young at the time, but needless to say, he didn't take up her offer. That was the last time I saw Grandpa; there was too much bitterness in the family and he never came again.

The rain driving against the window brings me back to the present, and the sound of rainwater, running from the roof guttering into the catchment butts, comes faintly to my ears. The familiar sounds of the old house have a personality of their own. There are creaks of protest as gusts of wind batter the aging walls and roof. The sounds are reassuring for they mean home to me.

I guess Mother's right; I have changed, I think as I pull the bedsheets up. *After all, I'm seventeen and I've had to grow up fast these last few months. But then, so has she changed since Dad died, and that's not long ago.* Looking back on the first few weeks of my leave, I can see the difference in her. *She's living in a world of her own, a world peopled by the past and centred around the small pub on the hill overlooking the forest and farmlands where we live.*

At times I don't think the present registers in her mind; maybe that is why she asked me to take her to see a film on the war at sea in the local movie theatre. It was a disaster. My nerves were on edge after my recent experiences in the Atlantic, and I went to pieces when the first ship was torpedoed. In uniform and with tears streaming down my face, I hurriedly left the theatre. Mother followed behind me up the aisle. We walked home across the public golf course and through the forest. "I . . . I'm sorry, Gordon," she said, blaming herself. "The . . . the nightmares. I hear you at night sometimes talking to yourself, but I didn't think."

"Never mind, Mum," I'd replied, "it's just that at the moment it's too fresh in my mind, the memories I mean. I just can't take it."

I turn over in bed, pulling the covers with me. The fire in the grate no longer flickers; there are just red-hot embers. Sleep comes at last; it creeps up, blurring my thoughts, and I fall into a jumble of dreams.

My suitcases lie open on the bed as I pack my worldly possessions, including the talisman. It's two months since I came home on leave. Without the companionship of John, the streets and theatres of the City of London and Soho have lost their attraction. I'm restless, ready to go back to the companionship of the sea. The company telegram is vague, just the ship's name, *Empire Harmony*, the dock number and port, West Hartlepool, plus general routing instructions, and the words *oceangoing*. She's not a coaster. A smile lights my face. I've been promoted from third to second radio officer, and now there is a second wavy gold band sewn on my uniform sleeve.

It is mid-March, the month of winds and rain, when I leave London with its nightclubs and theatres that never close, despite the nightly bombing. I settle back in my seat as the train heads north. I'm on my way and a feeling of excitement edged with uncertainty surges through me while I watch the passing countryside flash past the train's windows. I wonder when I'll see London and my family again.

The talisman is in one of the cases above me. It is an enigma, a symbol of superstition, like a lucky charm to ward off evil spirits. For whatever reason, I can't, by my own act, abandon the talisman. Deep in me lies a superstitious fear, a certainty that if I destroy the talisman, a greater harm will come to me.

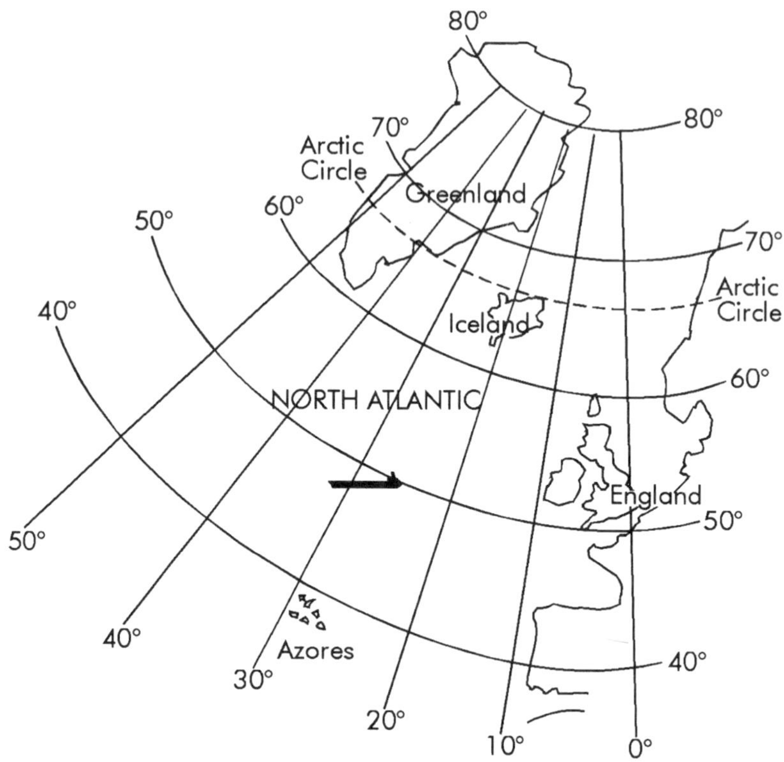

Black Pit, showing position of *Scottish Heather*

PART TWO

The Rock of Gibraltar

DISTORTED BY THE HEAT HAZE, the distant African coast dances in the binoculars despite my efforts to hold them firm. Scanning the horizon on the for'ard starboard quarter, I adjust the focal length slowly, and the coast zooms closer. The sandy reddish-yellow coastline looks barren. There are low-lying hills inland, but the shore is just sand and rock.

A voice disturbs my thoughts. I recognize the north country accent; it's the first officer, more familiarly known by the seagoing fraternity as the mate. "Not much to see there, Sparks. That's Morocco; it's desert." He pauses, then points towards the bow. "Look for'ard! A couple of points to port off the bow. See that smudge on the horizon? That's Gibraltar."

The mate stands beside me as I lean on the bridge bulwark, my elbows resting on the smooth wooden moulding along its upper edge. Shifting the glasses, I focus them on the distant smudge. "Yes, I see it," I exclaim excitedly. In the distance Gibraltar rises like a humpback whale, jutting out hundreds of feet above the sea.

His grey-blue eyes twinkle as I ply him with questions. "Come into the chartroom. I'll show you the chart." I follow him into the wheelhouse. The helmsman does not look round; his eyes are watching the compass and the bow, anticipating any deviation from the course. The wheel moves smoothly when he turns it, adjusting for drift.

The mate is fair-haired and tall with a rather large nose and friendly eyes, and an easy manner. He treats the third sparks and me like a couple of kids. I like him; he reminds me of the second mate on the *Scottish Heather*. A week ago, he showed me how to shoot the sun with the sextant and work out the ship's position.

The *Empire Harmony* is approaching the Straits of Gibraltar. Three weeks have passed since we left the ship's home port of West Hartlepool, bound for operations in the Mediterranean. That was on March 28; now it is mid-April of 1943. The convoy has slipped through the Atlantic undetected by the enemy. We'd kept far out to sea, heading west for days, and then south before wheeling to the east.

This morning I came off watch at 0800 hours and am not due on again until late this afternoon. We should be in Gibraltar by then. I like it here on the bridge in my off-duty hours, especially now the weather is warm.

The chartroom is behind the helmsman. The mate lays out the chart of the Straits of Gibraltar and, with his pencil point, traces the course to a cross just outside the Bay of Gibraltar. "We'll arrive here this afternoon about 1500 hours, and take a pilot aboard to enter the anchorage. Now, look how narrow the Strait is." With a pair of calipers he measures off the distance on the chart. "See, it's only about nine miles at the narrowest point between Gibraltar and the coast of Morocco. There's an old saying, *Whoever controls Gibraltar, controls the Strait.* This place has been fought over since Roman times. The British captured it from the Spanish in the early 1700s, and we've been here ever since."

Using his pencil as a pointer, he backtracks along the course to our noon position. "Right now, we're roughly here." The mate eyes me quizzically. "Any more questions?" I shake my head. He chuckles, "Okay, then. Let's see what's happening outside." I follow him out onto the bridge.

We've changed into more suitable clothes for the warmer weather. The navy blue uniforms have been replaced by khaki shirts and shorts with matching knee-length socks and brown shoes. Black shoulder epaulets, showing gold ribbon rank stripes, are slipped onto the shoulder straps, and a white cover is fitted over the crown of the high peaked officer's cap.

The convoy has entered the Strait. The coasts of Morocco and Spain are close now on our starboard and port for'ard quarters. Gibraltar lies ten nautical miles ahead, and I can make out more details. Light glinting off glass windows and the lighthouse on its most southerly point are plainly visible through the binoculars. The landward face of the Rock is sheer, and there are bare flat surfaces on its steep, sloping seaward side. They look like ugly, man-made scars.

"What are those large rectangular patches on the rock face for?" I ask the mate as he comes out of the wheelhouse.

He pauses briefly and glances at the Rock. The patches are plainly visible without the aid of the glasses. "Oh, they're concrete water catchments. When it rains, they drain the water into large underground reservoirs. They've been there many years." He hurries down the ladder to speak to the bosun on the quarter deck.

It's mid-afternoon. Gibraltar lies on our port for'ard quarter as the convoy slows and heaves to. Signal flags are fluttering on the commodore ship's mainmast. An Aldis signal lamp chatters and now there's a swarm of activity amongst the deck crew. Our signalling halyard sprouts flags of different shapes and colours, giving everything from our convoy position number to a request for a pilot. Our section is leaving the Algiers-bound convoy, and entering Gibraltar for further operational orders. The captain's on the bridge now with the mate.

The ship's telegraph rings for *slow ahead*, and, in company with other ships, we turn to port. The convoy sails on to Algiers, leaving us outside the boom and minefields protecting the anchorage in the Bay of Gibraltar.

The Rock stands firm and solid like a protective mother, its vast bulk huge and impressive on our starboard bow. With the pilot aboard, we enter the bay; the boom ship closes the entrance when the last vessel passes through. The *Empire Harmony* is now anchored in the safety of the Rock of Gibraltar. Under the protection of the Rock's massive armaments, airfield and protective minefields, we feel safe.

The mate, Jimmy, Ralph and I are sitting on deck chairs in a semicircle on the boat deck, chatting and enjoying the quiet of the night, free from the noise of the now silenced ship's engine. The small table in front of us is littered with glasses and bottles of beer. The large hump of the Rock rises to one side of the ship and, across the bay, the lights of La Linea flicker on its calm dark surface. It is peaceful and pleasantly warm; the night sky is clear and lit by the light of innumerable stars. A feeling of contentment settles over me.

Jimmy's face is animated as he talks. The dim rays of a shaded light flicker across his florid features. He's the chief radio officer and heads our section. Of medium height and in his mid-thirties, he is holding forth on his favourite subject, women. "Aye," he says, "I was once a projectionist in a cinema in London, you know. You'd be surprised what goes on in cinemas." His soft Scottish burr launches into one of his many exploits.

I like Jimmy, he's easy to get along with and we all call him by his first name. He and the mate are close friends. I used to call the mate *Mr. Mate*, but now that's changed. Ralph and I call him *Chief*; it's not so formal. Ralph's the third and I'm the second radio officer; we share a cabin on the boat deck.

Like me, Ralph's young, probably eighteen or nineteen. At just under six feet, he's a couple of inches taller than I. We're both about 140 to 145 pounds, so there's not much spare flesh on either of us. Ralph's got a thin pointed face and high cheekbones. He doesn't say much and keeps to himself. His black hair is slicked back and, unless he shaves twice a day, the dark stubble of his beard shows up against his sallow complexion.

The high-pitched hum of an approaching motor torpedo boat (MTB) attracts our attention and we idly watch it pass between us and another anchored merchant ship. The sudden explosion of a depth charge startles us, and we line the rails looking at the MTB as she disappears in the night. A white column of water sprouts up in her wake followed by the noise of another detonating depth charge. The detonation sends little shock waves vibrating through the ship, rudely shattering our feeling of safety.

The mate shrugs and, with a short laugh, says, "Well, that's put paid to the peace and quiet around here." He sinks back again in his chair, and drinks his beer.

"Why are they firing depth charges, Jimmy?" I ask. "Won't they spring the plates in the ships' hulls?"

"Ask the mate, Gord. He'll tell you."

The mate's eyes are fixed on his beer, and a wry expression spreads across his face as he says, "Well, they're certainly not fishing, that's for sure." He laughs and continues, "No, they are on anti-frogman patrol. Those depth charges are small anti-personnel types. They'll kill anyone swimming near them, but the explosive charge is not powerful enough to damage a ship's hull." His eyes follow the receding shape of the motor torpedo boat as he adds, "Unless, of course, they're set off very close to it."

The noise of the motor torpedo boat's engine, followed by the crump, crump, of exploding depth charges, continues periodically as it patrols the anchorage.

It is a phase of the war that is new to me. I listen intently to the mate and others as they discuss the menace.

"Aye," says the mate. "Spain's supposed to be neutral, but it's no secret that frogmen operate at will on this coast. Look at that shoreline! Miles of it, right from Algeciras to the border here at La Linea."

He nods his head towards the lights of Algeciras at the far western side of the large sweeping curve of the bay, then pausing, sips his beer before continuing. "They operate in teams from small boats at night outside the perimeter of the anchorage."

"Humph," grunts Jimmy. "They've got guts to do that, especially with those bloody patrol boats dropping depth charges."

"Yeah," the mate says, pausing to listen to the far-off explosions of the depth charges. "Damned if I'd like the job, but they do it." He gives a short sardonic laugh, then goes on. "Imagine swimming underwater with limpet mines strapped to your chest, and those bloody depth charges dropping near you. Bloody near certain death."

"Aye, man. Bugger that for a lark." Jimmy's voice adds to the conversation. "And, they've got to get to the ship, attach the bloody limpet mine, set the timing device and make their escape undetected. Christ," he says with a grin, "I hope our watchman's got good eyes; he'll have a hell of a job seeing any bubble trails at night."

In the few days we are in Gibraltar, several ships are put out of action by holes blown in their hulls. These ships are either beached or towed away for repairs. From the upper deck outside my cabin, I watch the motor torpedo boats on their nightly patrols. Judging by the noise of the exploding depth charges I do not envy the frogmen their job.

Shore leave! The highlight of a seaman's existence. After three weeks or more at sea, my thoughts turn invariably to the pleasures and excitement of the town. Shore leave gives me a chance to get off the ship and stretch my legs. No sooner had we anchored than the gangway was lowered, but it wasn't until today that the liberty boat was organized.

The deck officers run the boat on a rota system and a departure/arrival time schedule from the ship to the main quay has been set up. As I watch from the main deck, groups of seamen, dressed in casual clothes, call to each other as they board the liberty boat.

"Hi, Nobby," shouts a stoker, still wearing his sweat rag around his throat, "see you ashore later tonight in the pub." His friend on board, looking down on the motley crowd, laughs and waves back.

In the evening, lost in thought, I sit outside our cabin on the boat deck with Ralph. Small pinpoints of light, singly and in clusters, break the darkness of the hills on the Spanish mainland. There is a restlessness in my bones, and I want to discover what lies beyond the lights, to walk through the coarse dry grass and scrub covering the hills.

"That's San Roque, Ralph," I say, nodding towards the lights. "I checked the chart today. It's just a few kilometres inland from La Linea." I pause, my eyes searching his face. "We've got our passes so the border will be no problem. Do you fancy a walk tomorrow?"

He shakes his head. "No, not really. I've got some things I want to do in the morning. I'll probably go for the mail in the afternoon."

My enthusiasm undiminished, I make up my mind to take the first liberty boat ashore in the morning.

The next day I'm up early, and, washed and dressed, I have a hurried breakfast. Now, settling back in the liberty boat, I watch as a crewman casts off the painter holding us to the gangway. The third mate is at the controls, one hand on the gear change lever, the other on the tiller. "Off early today, Sparks," he says. "Where you going?"

I wait for the roar of the engine to settle down to a steady rhythm, before replying, "San Roque. It's up in the hills across the border. It'll make a nice change from the ship. Besides I need the exercise."

The third stretches his legs. Tilting his uniform cap to the back of his head, he chuckles, "Sure you're not after a bit of skirt, Sparks?" A grin spreads across his face. "Watch out for those girls in La Linea. If you're not back by tonight, we'll send a search party for you."

The thought amuses me. "Hadn't thought of that," I say with a laugh, then add, "Don't worry, I'll be back this afternoon."

A seaman on board an anchored merchantman waves and shouts to us as we round its stern. The dock is ahead and the third throttles down, then slams the gear into reverse as the prow of the boat edges alongside. I step on the landing stage and run up the steps to the dock. Our small group splits up and disperses as we each go our own way. The small garrison town of Gibraltar lies beyond the dock gates. Heading north towards the border, I pass through its quaint shopping centre.

The town lies at the foot of the western lower slopes that quickly rise, almost sheer in places, to the top of the rocky peninsula that forms Gibraltar. The peninsula is roughly three miles long and rises to a height of 1,350 feet at its highest and most southerly point. A narrow, low-lying isthmus joins Gibraltar to the Spanish mainland, across which runs the border and the Spanish town of La Linea.

A stretch of flat open land spreads out before me when I leave the town and reach the isthmus. The noise of aircraft engines revving up for takeoff breaks into my thoughts and I glance through the close mesh of a high barbed wire fence on my right. It is the military airport, and aircraft are already on the runways, taxiing or waiting for final clearance from the control tower. Only a short distance away, running parallel with the airfield, lies the border crossing point. Through this busy main gate stream Gibraltar's civilian work force of clerks and traders, visitors and army personnel.

The swarthy, thick-set border guard hardly glances at me as I cross into Spain. From my apparel it is evident who I am. With my peaked naval cap firmly on my head, I set off through the dusty little town of La Linea on the road to Algeciras. A kilometre or so along this road I turn inland on a dirt road leading to San Roque.

On closer inspection, the small picturesque houses in the village are cracked and dirty, the stucco walls uncared for and aging. In contrast, colourful blossoming tropical shrubs grow in ill-kept gardens. In the village, a small park stands alone on the crest of the hill overlooking the bay.

In the park, neatness and order prevail. Small flowering plants and shrubs in stone-edged, raised flower beds line the sides, while trees with large overhanging branches and leaves provide shade for the wooden benches beneath them. Rectangular in shape and paved with flagstones, the park is empty of people except for a couple of armed Spanish policemen. They smile and nod as I enter the park, but do not speak.

There's a light warm breeze here up high in the hills. It ruffles my hair when I take off my uniform cap and place it on the seat beside me. The bay is far below and I can see the ships in the anchorage, uniformly spaced in a large rectangular pattern in the northeastern quarter. Across the bay at its westerly end is Algeciras, a small Spanish port.

I try to pick out the *Empire Harmony*, but, without binoculars, the task is hopeless. The ships are stationary, indistinct blobs on the surface of the bay. My thoughts wander as I half-doze in the shade of the overhanging tree.

Unconsciously, memories of the *Soborg*, an old Danish collier, pass through my mind. The crew were all Danes except the chief radio officer and myself. As I think about her, my hand brushes the hair back from my forehead in the shifting breeze. It seems an age ago, yet only five months have passed since I sailed on her.

The sun is warm on my face, and I laugh and stretch my arms as I loll back on the wooden bench. The sun of Spain and the Mediterranean. Yes, I prefer these sunny climes to the harsh, bitterly cold weather of the North Atlantic. I'll never forget the winter storms between the northern coast of Scotland and the Orkneys. No wonder the Scottish coast is so rugged. Nor will I forget the thick crusted ice on the ship's rigging and decks off the Icelandic coast.

These and other memories bring a smile to my face as I get up from the seat. Setting off down the dusty path, I have a vision of the thick, choking coal dust swirling over the for'ard cargo decks as we bunkered a Murmansk-bound convoy in Hvalfjord, a fjord in Iceland. I was helping the deck officers tally the bucketloads of coal being hoisted out of the ship's hold.

God, that was a barren and inhospitable place. Located at the foot of a steep cliff and perched on the bare, rock-strewn shore of the fjord, the village was just a collection of stone houses and a general store with a fishing boat jetty. Poor as it was, however, it was a milestone for me, because it was the first time that I had trodden on foreign soil.

The grass is sunburnt and brittle on the hillside. The dry stalks crunch beneath my feet and seeds cling to my knee socks. I stop and pull the seeds off when I reach a footpath. Worn bare by countless feet, the path runs diagonally down the sloping land to the Algeciras-La Linea road below.

Jaunita

"THEY'RE NICE, aren't they?" she says as I admire the miniature Toledo swords in the curio shop window. The voice is soft and feminine. Startled, I turn round and find myself staring into the smiling eyes of an attractive young woman.

"Wh' . . . well, yes. I was just window shopping," I stammer, flustered and not quite sure what to say. Then turning back to the window, I gesture with my hand towards the small Toledo swords. "They're beautiful. Look at the fine filigree work on the sword handles and the guards." I shake my head regretfully. "They look expensive to me."

"Most of the curios are expensive on the main street. They're cheaper in the side streets. You're off one of the ships in the bay, aren't you?" she asks.

"Yes, I've never been here before, so I'm just sightseeing."

"Would you like me to show you around our town?" Her hand takes mine and her dark brown eyes twinkle. "It'll be company for me, too." She has on a light blue summer dress with a low neckline which shows the deep cleavage between her breasts. She's well developed and the dress leaves little to the imagination, for she is not wearing a bra.

Flattered and impressionable, how could I not agree. She's a beautiful woman. "I would like that very much," I say and feel the responding pressure of her hand.

She smiles and runs the fingers of her free hand through her long black silky hair. We walk, hand in hand down the main street. "What's your name?" I ask her.

"Juanita. You like that name?"

"Yes, it's a lovely name. My name's Gordon."

We turn down a quiet side street, and into another as she leads me to a small back-street antique shop. The door is open, and I blink my eyes as I follow Juanita in. The interior of the store is dim in contrast to the strong sunlight outside. "This is my uncle's store," she says. "He deals in antiques and curios. I'll try and get you a good bargain for those Toledo knives."

She's talking in Spanish to the old man behind the counter, and calls me over as he fumbles in a drawer. "Aaahh, yes," he says, "I still have some left." He lays the miniatures on the counter.

The old man talks as I inspect them. "Toledo," he says nodding his greying head, "used to be the capital of Spain, but that's a long time ago. It's on the Tagus River about forty miles south of Madrid." He picks up one of the miniature swords on the counter, and examines it. "Now, look at the intricacies of the workmanship in the hilts and hand guards. They are cheap, too, considering the silver filigree work."

The swords are six or eight inches long, and delicate. Undecided, I look at Juanita as she talks to the old man. He smiles and says, "For you I will make a special price."

With Juanita's help in bargaining, I buy one of the miniatures. There's a twinkle in the old man's eye as he parts with the souvenir. "Bring your friends back next time you're here," he says, "and I will give you a special discount."

Smiling my thanks, I assure him that I will. With the souvenir in one hand and Juanita's small hand clutched in the other, we wander out into the bright sunlight of the street.

"Is he really your uncle, Juanita?" I ask, as we pass down the street.

"Well, not really; but he's an old friend. I've known him for years, and think of him as an uncle." She glances up at me, her brown eyes looking into mine. "Don't you also have the same thing in England?"

"Yes," I nod and add, "I've several uncles who are not related by blood. They're just friends of the family who have known me since childhood."

Juanita pauses at the corner of the street as if undecided on which way to go, then looks up at me as she announces casually, "I live quite close to here. Would you like to come home with me?" It comes out quite naturally, almost as an afterthought.

I look at my watch. It's still early in the afternoon. I hesitate, then say, "Why, I'd love to; I've plenty of time. The liberty boat leaves the main quay at five o'clock, then once an hour after that."

She laughs. "Come on then, let's go." She tugs my arm and I feel her hand tighten in mine. Something in her manner arouses me, and I become aware of her sexually as she puts her arm around my waist. It is a new experience for me, and I feel a sense of excitement and the unknown. I realize belatedly that this is probably no chance encounter.

My arm is around her shoulder as we walk through the back streets. There is a subtle change in the character of the neighbourhood. The houses are becoming seedier and I am aware of eyes following us, eyes which bore into my back as I walk with Juanita. There are no sidewalks, just dusty streets. Little groups of children play in the dirt, while people sit in doorways, watching the world and their neighbours going by.

Juanita points to a flat-roofed house across the street. It is a single-storey building with whitewashed walls, one of many in the long row of similar style houses. "Come, we're home now. Let's go in." She pushes open the door and I follow her into the room.

Juanita murmurs something to me, something about a madam, but I'm not taking it in because my mind panics. *How did I get into this mess?* I ask myself in despair as full realization hits me. Juanita's home is a brothel.

All hidden thoughts of sex have gone as Juanita leaves me in the room. I'm alone and afraid, not physically afraid, but afraid of myself and embarrassment. I feel as if my very being is about to be stripped naked. *What do I want? Why didn't I just say, calmly and quietly, No, I'm sorry, Juanita, maybe another time.* It would have been so easy.

Now as I wait in the small reception room for the madam, these and many other thoughts surge through my agitated mind. *"Don't go with dirty women, Gordon,"* my mother had said, as she painted word pictures of the horrors of venereal diseases. *"Save yourself for your future wife."*

I'm sitting at a polished table in a furnished room. There's a window behind me, I noticed it when I came in. It's got pink curtains drawn back to let the light in. The sun's rays reflect off the polished floor and cast pools of light on the ceiling, which, like the walls, is painted a subdued cream. A settee with a small patterned rug in front of it occupies one wall, while armchairs are arranged tastefully around the room.

It seems an age, but in reality it is probably only a minute or so before an inner door opens. The madam, an elderly woman, smiles when she sees me. "Do you want to go with Juanita?" she says. Then, when I don't answer right away, she adds. "We do have other girls here if you want someone else."

"No, I don't want someone else." I panic, saying the first thing that comes into my mind.

"Then what do you want?" she says. "You want to see a girl's breasts?"

"No." Shame and humiliation overcome me. I wish the floor would open up and swallow me, as I'm forced to admit the truth. "No, I'm sorry, but I don't want to go with anybody."

The wrinkled face of the elderly lady changes as she smiles at my embarrassment, and the crow's-feet at the edges of her eyes deepen. "How old are you?" she says in a gentle voice.

"I'm seventeen."

"Seventeen? You're much too young to be in the war."

My pride is hurt. "I'll be eighteen soon," I reply.

She laughs softly and shakes her head. "I think if you want to come again, it would be better if you brought some of your friends with you for company."

The madam escorts me to the door and points the way to the border gate. A weight lifts from my mind as I hurry down the dusty street. The sun feels good on my face, and the escape from imagined dire things to come spurs my feet onwards. I walk quickly through the winding streets of La Linea to the border crossing, through the checkpoint and into Gibraltar.

The liberty boat drifts idly on its painter tied to the landing stage. There's still fifteen minutes to go before the scheduled departure time. I join the third mate as he sits on a bollard near the boat. At twenty-seven, he's a hardened seaman and no stranger to brothels. He laughs when I tell him about my encounter with Juanita. "Christ, Sparks," he says, as he pushes his hand through his fair, swept-back hair. "You should have fucked the bitch. What's the matter with you. Aaahh, well, you'll learn."

He stops talking. Returning crew members are approaching and he glances at his watch. "It's time to get back. Let's get aboard." We wait five minutes for latecomers and then start back across the water to the ship. It's late afternoon and the sun is beginning to dip towards the west.

The familiar sight of our ship, the *Empire Harmony*, comes into view. She is gently swinging on her anchor as the liberty boat approaches her gangway slung down the port side. With those massive jumbo derricks you cannot mistake her. She is a crane ship, ideally suited for work as both a general cargo carrier and, more importantly, as a heavy lift ship to replace destroyed dockyard crane installations in the ports of war-torn Europe and North Africa.

The nightly chorus of depth charges forms a background to my thoughts as I lie awake in my bunk. I listen as the harsh roar of the motor torpedo boat fades in the distance, carrying with it its deadly cargo.

I'm restless, and sleep will not come. *Why? Why do I panic and run blindly when sex rears its head? What must Juanita think of me?* I hear again her voice; feel once more the torment of shame and degradation at my panic-stricken response to the madam. I lie on my back staring up at the deck head as I try to grapple with my feelings. The events of this afternoon are still fresh in my mind. I can't understand the fear within me, the fear of women. I envy the easy casual manner of Jimmy, my chief, and others like him, envy them for their love affairs and conquests.

I'm well into my teens now, seventeen, for Christ's sake. I'm a growing man with all the sexual desires that go with it. Bitterness eats at my soul as I search my mind trying to understand these ingrained responses, and curse their cause. I feel like a clumsy country yokel when it comes to women.

Memories of my early childhood parade through my mind. I remember the nuns of St. Mary's with their thick leather belts. I remember those belts well; they used them often enough on me. I laugh bitterly at the thought. *Did they start off all these fantasies that plague me, sexual fantasies?* Their faces rise before me. Sister Philamena was hard and merciless. She and others bound my hands behind my back and beat me with those belts in front of the class. Those beatings kindled my sexuality, an unnatural sexuality which tormented me, a sensuality engendered by pain and guilt.

The nightmares of my childhood have never left me. They occur less often these days, but there's a fear, a very real fear which amounts almost to panic at the thought of my hands being trapped behind my back. Just thinking about it makes me break into a sweat, and I tear myself away from the thought. *Is this the legacy of the nuns of St. Mary's, or their counterparts, the brothers of St. Egbert's?*

My younger brothers had seen the nuns beat me and told my mother. Now as I toss and turn in my bunk, I ask myself, *Why, why didn't she stop them?* It was soon after that first beating that the fantasies and nightmares started. Now I don't know what to think, except that I began to hate the nuns, and, later, the brothers of St. Egbert's.

The patrol boat has returned. I can hear its powerful engine as it passes, and wait expectantly for the crump of the depth charge, and the shudder as it hits the ship's hull. It comes and passes, the sound of explosions dying away as the patrol boat moves further down the lines of anchored ships.

I turn over and try to sleep, but sleep is slow in coming. My agitated brain holds court on Mother's influence over me. Her power was total. I even had to tell her every time I went to the toilet. That started when Mother asked me about masturbation, and gave me a lecture on its evils. Strange, but for years in my childhood and early teens, the habit of telling her persisted. *Yes,* I thought, *Mother taught me well, and instilled in my young mind a terrible feeling of shame and guilt. Sex was a dirty and nasty thing. There seemed no room for love in her life.*

It's as if I've been brainwashed, conditioned by the nuns and my mother to follow their dictates, just as I did this afternoon when I panicked and fled.

I don't go back across the border to La Linea. I play it safe and stay away. Instead, Ralph and I wander the old town exploring the quaint winding cobbled streets and old-fashioned stores; that is, until we discover Catalina Bay on the other side of the Rock. Single-decker buses run daily from the town square to the bay.

"Come on, Gord," Ralph says when we look at the bus timetable. "Let's go." He points to a bus standing at the terminal. "That's got to be the bus."

Around the southerly tip of the Rock, which juts out into the Straits, the view is spectacular. The Mediterranean, stretching to the east, is a vast expanse of blue squeezing into the bottleneck of the Straits of Gibraltar and out into the Atlantic on the west. The coast of Morocco is clearly visible. Then as we round the point, it is hidden by the turn of the road.

The bus changes gear as the road steepens and follows the base of the towering cliff side. The road enters a tunnel cut through the limestone before emerging on the eastern face of the Rock. Above us, the large concrete water catchments cover the steep slopes. There below us as the bus emerges from the tunnel, Catalina Bay stretches out in front of us to merge with the distant Mediterranean.

The road, cut into solid rock face, slopes down to a narrow strip of sandy beach. We have arrived. Ralph laughs and says, "We should have brought our bathing suits, Gord. We could have gone for a swim."

"Ah, well," I say, "maybe another day." I glance at my watch and nod towards one end of the beach. There's a small beach restaurant with a row of tables, each with a colourful umbrella. "It's lunchtime; let's get something to eat and a cool drink." I take off my uniform cap and wipe the sweat from my forehead, "It's bloody hot, Ralph." I pause to look around, then gesture with my hand. "Over there in the shade looks nice."

Seated at one of the tables, we watch the couples sunbathing on the beach and swimming in the sea. Men far outnumber women in this outpost of the Empire. It would be strange if they didn't, because Gibraltar is an armed camp.

Our ship lies at anchor in the bay for about a week before we receive our orders to join an eastbound convoy going through the Straits of Gibraltar to Algiers. The opportunity to visit Catalina beach again passes by. Even my eighteenth birthday goes unnoticed and uncelebrated, but perhaps that is my own fault because I forgot it myself.

Malta Convoy

THE RATTLE and hiss of steam from the foc'sle winch, followed by the clang of the heavy anchor chain being pulled through the bow port, wake me. It's first light and I can hear the mate's voice issuing orders to the crew. The anchor's being raised and the *Empire Harmony* readied for sea. I look at my watch. It's six in the morning, late April 1943, and I'll be on watch in half an hour. Jimmy takes over from me at 0800 hours. Our orders are to sail for Algiers where we will be given our final destination at a naval conference.

Slipping out of my bunk, I wash and dress hurriedly, then wander out on deck. The pilot is aboard, and I hear the engine room telegraph ring, then feel the deck vibrate beneath my feet as the propellers bite into the still waters of the anchorage. The Rock, bathed in the early morning sun, lies sleeping on the port bow. Taking a last look, I breathe in the fresh morning air before turning to go on watch. We are on our way.

The *Empire Harmony* joins the small line of vessels threading their way through the boom and the minefields beyond. After dropping the pilot into a waiting launch, the ship surges forward and, gathering speed, heads for the waiting eastbound convoy in the straits.

It's pleasant here, leaning against the ship's rail, idly watching the lines of ships ploughing through the calm blue seas. The morning sun is warm, and I can feel its rays through my short-sleeved khaki shirt. Jimmy's taken over the watch and the war seems distant and remote as I absently study the ship on our port beam. There's a hurricane fighter mounted on a metal ramp on the foredeck. I've never seen a merchant ship with an uncrated aircraft on her deck, especially one on a ramp.

Suddenly, without warning, a violent explosion rends the air, hurling debris and smoke into the sky. The strange-looking ship, her back broken, settles in the water as the shock wave hits us like a hammer. Instinctively, I crouch down, shielding my face. The *Harmony* heels over, then sways back again as the wave passes. The explosion brings all hands scurrying out on deck like ants from a disturbed nest.

Rooted to the spot, I watch the death throes of the ship as she begins to break in half. The escorting naval vessels split; two comb the sea on the port side for U-boats, while the third closes in on the sinking vessel to pick up the crew.

I can see the tiny figures of men on her decks as they abandon ship. With the initial shock of the explosion passing, I stand there, my body tense and hands gripping the rail, a helpless spectator. Fascinated with the unfolding drama, my mind urges on the efforts of a pilot as he gets into the cockpit of the hurricane fighter. Puzzled, I turn to a crewman standing next to me and say, "He's going to try and fly that aircraft off the ship. What type of ship is that?"

"Oh, aye, Sparks," he says. "That there ship's a CAM ship. It's short for Catapult Aircraft Merchantman. The trouble with them is when he's out of range of land, the poor bastard has to ditch in the sea near the escort." His eyes return to the scene as he adds, "With luck, he may be able to take off but the ship's going fast."

Our eyes are riveted on the aircraft as the pilot tries to launch the fighter. A strained sigh of relief ripples through the group of our crewmen as the aircraft flies free, leaving a stream of white smoke or steam behind. Circling the convoy, the hurricane flies low over the doomed ship, then heads off towards the nearest airfield.

"There," the crewman turns to me, grinning in triumph. "He's made it. That man's lucky. He'll fly to Gibraltar and land there."

We cannot stop, but sail by the sinking ship, leaving her behind as she founders in the sea. Her crew are picked up by the escort and the convoy continues, without pause, eastward along the coast.

An hour or so later, a second ship is torpedoed by a U-boat which has penetrated the naval escort screen astern of the convoy. The convoy, at maximum alert, keeps steady on course, maintaining a speed of eight knots. We should be in Algiers in two days' time if all goes well.

On schedule, early in the afternoon we see the city of Algiers ahead; there have been no further attacks. Moving into the protected anchorage, we drop anchor. The mate comes down from the bridge. "Bosun," he calls to a burly seaman in a group of sailors near the foc'sle head. "Get some men and lower the gangway."

Turning to his men, the bosun speaks quickly, giving orders. They move as a group across the main deck to the midships accommodation close to where I stand. Interested, I watch as they remove an opening in the outer bulwark of the ship.

Stored on its side, the gangway lies folded horizontally against the ship, locked into the superstructure of the main deck accommodation. A jib, an integral part of the gangway mechanism, swings out and the gangway unfolds. The crew secure the top end to an opening in the ship's bulwarks. To a crew member operating a block and tackle attached to the gangway's lower end, the bosun says, "Okay, Ron, lower away." He watches from the top of the gangway, gauging the lower platform's height above sea level. Raising his hand, he yells, "Stop!" He looks briefly at it, then says, "Okay, make it fast." The small square platform at the end of the steps is about two or three feet above the waves.

A naval launch is heading towards us, pennant flying in the wind created by her movement, her brass attachments gleaming. A rating in white tropical uniform stands on the deck with a boathook in his hand. The helmsman throttles down the engine, and with a precision gained from long experience, draws up alongside the foot of our gangway.

Forewarned, our captain, his face set, hurries down and boards the waiting launch. With a burst of power from the engine, the launch takes off, racing across the water to other ships in the anchorage. Her mission is to collect all the captains for a naval conference ashore.

It is late when the captain returns and joins us at dinner in the wardroom. The naval conference is over and we have our orders. There's a gleam in his eyes as if he's enjoying keeping us in suspense. "Well," he says, his deep-set eyes glancing round the table. "I can tell you now, there's no shore leave for anyone. We're going to Malta. We will be here three days, then move up to Bône on the Algerian and Tunisian border." He takes a sip of water before continuing, "Once Cape Bon has fallen to the Allied Forces we sail for Malta."

The silence is broken by a hubbub of noise as talk breaks out around the table. The captain smiles, and his thick bushy eyebrows move up and down expressively as he answers questions. "Yes, Cape Bon should fall within a week." To another question he answers, "No, shore leave is out of the question for security reasons."

From the anchorage, the town looks inviting and mysterious. My mind conjures up romantic visions of Charles Boyer and stories of the French Foreign Legion. It's a pity all shore leave has been cancelled; I would have loved to visit the Casbah. True to the captain's word, three days later there is a burst of activity in the anchorage. Naval launches bustle to and from the ships which are raising steam for sailing.

One by one, accompanied by the hiss and rattling of steam winches, the vessels haul up their anchors and head out to sea, seven merchant ships in all. Formed into a convoy and protected by a heavy escort of an equal number of naval vessels, we head for the small port of Bône on the Algerian/Tunisian border, roughly fifty miles or so from the front lines, where the Afrika Korps holds a bridgehead around the town of Tunis and Cape Bon.

It is late the following day, May 11, when we anchor in a bay to the east of the port, having covered the three-hundred-mile voyage without incident. The convoy and escort lie at anchor waiting for Cape Bon to fall before making a dash for Malta.

Time hangs heavily on our hands. Speculation is rife on board ship concerning our chances of getting through. After all, there is a heavily armed escort of seven naval ships guarding an equal number of merchant ships. Someone in command must be expecting trouble.

In our off-duty moments, Ralph and I occasionally join Jimmy and the mate in Jimmy's cabin, or sit on chairs outside on the deck to discuss the topic on everyone's mind. Today is no different. The mate smokes his pipe, and lies back in his deck chair with his feet propped up. He is informally dressed in a khaki shirt and a pair of shorts. We chat amongst ourselves about the war news and other things.

Inevitably, the topic changes; Jimmy's spinning another tale about his adventures with women lodgers in the house where he stayed in London. According to him, he's slept with every woman in that house. "Aye," he says, his dry sense of humour and Scottish accent holding his audience. "'Tis, true ye ken. I wouldna kid you." Like most Scots, he has a knack for storytelling.

It is peaceful and relaxing sitting here, idly watching the black silhouettes of the rugged mountain ranges against the starlit night sky. *How long will this peace last?* Only the future knows, but for now we can only discuss the war news as we know it, news gleaned from short-wave radio broadcasts over the Armed Forces Radio and from other sources; some reliable, others not so. The North African Campaign, Operation Torch, is almost over. The joint British and American forces are in control of the land from Morocco to the Tunisian border. The Eighth Army under Montgomery, better known as the Desert Rats, have pushed the Afrika Korps back from Alamein on the Egyptian border to the outskirts of Tunis, the capital city.

"Yes," says the mate in reply to a question. "The last convoy from the west to reach Malta was in August last year. It was a heavily armed convoy with cruisers and destroyers for escort, and consisted of fourteen fast merchant ships."

"Ah, yes, I remember," says Jimmy. "The tanker *Ohio* was badly damaged, but made it despite being torpedoed."

The mate nods, "Yes, that's right. The cost in ships, both naval and merchantmen, was horrific. Only the tanker *Ohio* and two or three other cargo ships in the convoy got through; the rest were sunk." He pauses a moment to knock out the ash from his pipe and replace it in his shirt pocket. "That convoy raised the siege of Malta," he says, "and brought in desperately needed fuel and food, but the sea route to Malta from the west was abandoned until now. Once Cape Bon falls, our convoy will be the first to run the gauntlet from the west since then."

The mate reaches for a half empty beer bottle and pours its contents into his glass. "It's different this time though," he says. "The Germans and Italians have only Pantelleria and Lampedusa airfields and those are bound to be heavily bombed. True, there are other airfields in Sicily and Italy, but I think the Germans are too heavily engaged on the Russian front to do much about us. At least let's hope so."

"Yes," Jimmy agrees. "The situation's changed drastically since those days. With the loss of North Africa, the importance of Malta is no longer a real priority to the Germans."

With the fall of Tunis and Cape Bon on May 15, 1943, Rommel's Afrika Korps surrenders, and the convoy is ordered to sail to Malta. The island lies roughly one hundred and ten miles south of Sicily and about two hundred and eighty miles to the east of us. We have to pass close to the islands of Pantelleria and Lampedusa. The main threat to the convoy will come from enemy aircraft based on these islands and in Sicily.

With our heavily armed escort, the convoy leaves the bay early in the afternoon and sets a course for Malta. The large massif of Cape Bon looms up on our port bow as we pass its barren shoreline. Local fishing boats ply their trade, seemingly oblivious to the historic events taking place on their shore; the daily needs of their lives must be met. Naval patrol boats move among them, checking for escaping enemy troops. Jimmy points with his arm. "Look, Gord! Over there," he says as he hands me his binoculars. "That corvette, she's stopped one of the boats, and is sending off a boarding party."

Adjusting the binoculars, I can see the naval ratings under the command of an officer board the fishing boat. He disappears into the wheelhouse while his men stand guard, their guns at the ready. "Not much chance of any Germans escaping these patrols, Jimmy," I say with a laugh, handing them back.

Jimmy nods, "At least, not in daylight. They'd be better to wait for nightfall."

"Yeah, I guess so," I reply thoughtfully, "but then, it's probably a matter of opportunity. I suppose they'd rather risk the sea than a prisoner-of-war camp."

Cape Bon falls behind the convoy, and with it, the fishing boats. We're doing eight knots, about average for merchant ships of five thousand tons. I take over the radio watch from Ralph at 1600 hours. There are no messages for us or the convoy, just weather reports and long lists of messages for other convoys which I record in the log.

I'm off watch as we approach the island of Pantelleria, and now the ship is on maximum alert. We have army gunners (DEMS) aboard to man the Oerlikons on the wings of the bridge. The mate and the captain are in the chartroom. The army gunners in shorts and khaki shirts are lolling in the starboard gun nest. One is seated behind the Oerlikon with its double barrels pointed skyward ready for action, while his mate finishes checking the ammunition belt drive in the large canisters mounted on the gun itself. Their steel hats lie ready close by.

"Hi, Sparks," says the man seated behind the gun as I wander up onto the bridge. "Come to join us? We could do with a bit of company. Seems old Jerry doesn't want to tangle with us, aye, Bert?"

"Nah, Fred, I don't think he do, now. Must know we're bloody well 'ere, an' from what the corporal told us they've got airfields on that bloody island." I grin at Bert's accent. He's a real Southerner from the rural southern counties of Cornwall or Sussex, while Fred's a Londoner.

Bert is a short wiry man, brown as a berry. He's younger than Fred, somewhere in his early twenties, while Fred's in his thirties and tall. They're easy to talk to and I've known them since I came aboard.

"You checked that midship pop gun yet?" Fred asks, his grey eyes twinkling. "You reckon you can hit anything with it?"

"Well, I'll bloody well try," I say with a laugh. "It may not be powerful and have the range, like yours, but it's a Browning machine gun."

"Yeah, so you said before, mate. We know they use them in the hurricane fighters, eight of them. But now, this toy can fire shells, both incendiaries and armour piercing ones," Fred says as he strokes the breech of the Oerlikon lovingly. "True, Bert?"

Bert looks up, as he reaches for a packet of cigarettes and lights one. "Yeah, Oi reckon that's true enough," he says as he flicks the burnt-out match stub over the side. "But, you mustn't kid our Sparks, man. He's young and doing his best, is the lad, now."

"Here, Sparks," Bert says as he hands me his binoculars, "look at that smudge over there, 'tis Pantelleria, ain't it Fred!"

Fred nods as I take the binoculars and train them on the distant smudge. The island is no longer a smudge, and I can see the outline of its hills and mountain ranges. There is no doubt that they know we are here.

"Do you think they are waiting for night before they attack?" I ask, handing the binoculars back.

Fred shrugs his shoulders. "Hard to say. Depends if they're Jerries or Ities. Maybe the airforce has bombed 'em out of operation. If they don't come in the daylight, I think it's a safe bet that they'll come tonight."

We chat for awhile before I leave them to check the Browning amidships, just aft of the funnel and on top of the cabin I share with Ralph. It is my emergency post when we're under attack, while Jimmy's post is in the radio cabin. Metal rungs welded into the bulkhead give access to the machine gun nest, and, reaching them, I climb up onto the top of the cabin.

The steel-plated gun nest is surrounded by an additional layer of sandbags to provide a protective barricade for the Browning. The gun, with its belt drive canister, is set on swivelled mountings, giving it a wide arc of fire.

Removing the canvas cover to check the belt drive feed, I smile. It brings back memories of my years in the Air Cadets and Taffy, the armourer at the Northweald RAF Airfield. The air cadets were posted there in 1939 when I was fourteen. I used to follow Taffy around the hurricane fighters helping him arm and test the eight Browning machine guns in their wings.

He taught me to strip and clean those guns, and introduced me to his friends as his winger. "Aye," Taffy would say in that lilting Welsh accent I found so fascinating. "This is my winger. 'Tis Gordon you know." Those were happy carefree days, but the government withdrew the cadets from the airfields when the Germans started bombing them in the spring of 1940.

Now, satisfied the gun belt and mechanism are running smoothly, I climb down the steel rung ladder to the boat deck, and, after one last look, go down to the wardroom for dinner.

It is in the early hours of the morning, about 0100 hours, when the first air attack on the convoy is launched. We are halfway between the islands of Pantelleria and Lampedusa when the torpedo bombers come in low over the sea from the direction of Pantelleria. The shrill notes of the klaxon bring men tumbling out of their bunks and on deck or to their action stations.

I've been half expecting it, and am fully dressed, lying on top of my bunk. Grabbing my life jacket and steel helmet, I dash out to the roar of low-flying aircraft and the sound of gunfire from the escorting naval vessels. Turning to the steel ladder set into the outer bulkhead of our cabin, I scramble up to the machine gun nest.

Frantically, I look around for a target, the butt of the Browning clamped tight into my shoulder as I crouch in the nest, my heart pounding. There is no moon and I cannot see the aircraft coming out of the blackness of the night. They come and go in an instant, hugging the sea and launching their torpedoes.

Aimed low from a merchant ship close to us, tracer bullets streak across the sea. *What is he firing at?* I wonder, for I can make out nothing from our position in the convoy. I hear a yell from one of the gunners on the wing of the bridge. "Silly bloody fucker, he'll fucking well kill us, not the bloody Germans." It is frustrating sitting there straining your eyes into the dark looking for fast travelling aircraft flying at mast height directly at you. The planes are over and gone in seconds. *Was that just the first wave or are more waves coming over?* There were no bright flashes and explosions of hits on ships in that attack. I listen intently as quiet returns to the sea and the watching gunners. The convoy pounds steadily on towards Malta.

We wait for the next wave, but though there are false alarms and occasional gunfire, no further attacks develop. *Maybe Allied aircraft are bombing their airfields.* We hope so. The danger appears to have passed, because in the dawn light we see American twin-tailed P38 lightning fighter bombers streaking low over the sea close to the convoy, on their way to bomb and strafe the enemy airfields. I watch them, fascinated by their speed as they skim the wave tops. They look deadly.

The island of Malta comes into sight in the afternoon, and marks the end of our voyage. We round its coast and prepare to enter Valletta Harbour. One by one, the ships pass through the entrance to the harbour, watched by countless people who have one eye on us and the other on the sky watching for the inevitable air raids. A new phase has begun for Malta. With the Allies occupying the whole North African coast, they are no longer under siege.

Talking excitedly, Jimmy, Ralph and I lean against the ship's rail. "Christ, Gord," says Jimmy. "It's almost like we're liberators, man. Look at them, men, women and children. They're everywhere, even on the bloody breakwater, waving and shouting."

The excitement is contagious and brings a lump to my throat as we wave back. The *Empire Harmony* moves slowly down the Grand Harbour and berths alongside the main docks by the gateway to the city of Valletta, close to the battered and flattened suburb of Floriana.

We dock amidst a hubbub of noise from the excited crowds who watch from the tall cliffs which rear vertically behind the docks at that point in the harbour. Almost immediately gangs of swarthy men swarm aboard. Under orders from the mate, the stevedoring crews begin to remove the hatches and unload our cargo. The high explosives, ammunition, food and general cargo we have brought are all welcome additions to the island's needs, because their stocks are running dangerously low.

Valletta Harbour and the nearby suburbs have been heavily damaged by bombing. Now, with liberation of the North African continent and the lifting of the siege, the occupation of Malta is no longer a priority for the Axis powers. The enemy are more concerned with the defence of Sicily and the Italian mainland than with attacking Malta. The air attacks from Sicily and southern Italy continue to harass the Allied Forces on a reduced scale, and to slow the buildup of supplies for any future invasion of the mainland.

The sudden wail of air raid sirens interrupts work in the holds and on the deck of the ship. The labourers and stevedores scramble over the side of the ship and into the nearby air raid shelters, which are bored deep in the cliff.

The crew take up action stations and I scramble up the ladder to my post in the gun nest. Valetta's anti-aircraft guns open up on the high-flying enemy bombers, and their shell bursts pepper the sky. Frustrated, I sit behind the Browning, its butt firmly gripped against my shoulder, and realize the futility of my task. The aircraft are flying well above the range of a machine gun.

Watching the action, I feel secure behind the concrete and metal protective bulwark of the gun pit. It relieves my feelings just to sit there, alone, wearing a standard British tin hat to guard against falling shrapnel.

Due to the height at which the enemy bombers are flying, the bombing is very inaccurate, and many bombs fall wide of their target. The thin shrieks of the bomb tail fins increase in volume as they approach us, only to fall and explode in erupting water spouts just outside the harbour sea wall, missing the docks.

Not all of the bombs miss, however. The shrieking increases in intensity and strikes fear as the bombs explode, each one nearer than the last. They burst in line as the bombing run hits the town and dock areas, throwing up columns of smoke and debris. Some hit old ruins and re-align the smouldering rubble in grotesque fashions. Our ship, tucked up close to an overhanging cliff alongside the dock, is untouched during these air raids.

The nearby suburb of Floriana is familiar to us now. Sometimes I walk through the suburb by myself and sometimes with Jimmy and Ralph. The extent of the bomb damage surprises me. I remember Jimmy scratching his head in bewilderment and saying, “It’s like walking through a small town that’s been fought over by armies for months.” That suburb next to the docks and the other dock areas are virtually flattened.

I’m a loner, and most of the time walk the countryside on my own. I don’t know why I do this but I feel free and at peace with myself when I’m alone. The dried-up landscape, the dusty roads and countryside outside the city limits attract me like a magnet. Malta’s an island which seems to grow stones. The rough dirt roads and fields are bordered with walls, walls built of stone upon stone, the mothers, fathers and grandfathers of stones, with little baby stones filling in the spaces between the bigger stones, all arranged in terraces up the sloping hillsides.

It’s nearly two weeks since we arrived and we are sitting in deck chairs outside Jimmy’s cabin. The night is warm and the beer relaxing as we chat.

“Yes,” says the mate. “There’s a church built of human skulls here in Malta.”

“Oh, come on. That’s a bit tall to swallow,” I say with a laugh, my credulity stretched to the limit. “Where’s this?”

He smiles, unblinking, his eyes fixed on mine. “It’s true, isn’t it, Jimmy?”

Jimmy, not too sure, plays it safe. “Something to do with the Templar Knights, wasn’t it?”

Ralph says nothing. His eyes are locked in a fixed gaze on his beer. A listener, he doesn’t like to give an opinion.

“Well,” the mate continues. “The church is in Medina, a small town about a couple of miles outside Valletta on a hill. I visited Malta before the war when I was serving my apprenticeship, and saw the church. It really is built of human skulls. The history of the church goes back to the Templar Knights. The town of Medina is even more ancient. It goes back to Roman times, and possibly even to the Carthaginians. Now, if you want an interesting place to walk to, Gord, that’s the place. It’s not far from here.”

“You’re on, chief,” I say, my interest aroused.

The following morning finds me on the road to Medina. The dusty winding road is lined with stone walls and terraced cultivated fields, the height of the terracing varying with the steepness of the land. It’s midday when I reach Medina. I pause to look back on Valletta. It looks a long way. I sit on a flat rock by the roadside on the crest of the hill. There is a breeze up here and I can see the distant sea. The sun’s rays dry the sweat on my shirt as I eat my sandwiches.

As I wander through the town, I'm surprised because there's not the bomb damage here that there is in Valletta; but then, this town is of no military or industrial importance. The narrow and twisting streets widen out into large squares, and there are several churches. One church, more ancient than the others, is locked. Perhaps this is the one which the mate was talking about; but if it is built of skulls, they're not visible from the outside. I give up my search and retrace my steps through the town, down the road to Valletta and the ship.

In Valletta itself, life, at least for merchant seamen and their ilk, rotates around the bars and cafes of the *Gut*, a street of ill repute, well-known by seamen of all navies. Perhaps local inhabitants know it by another name. It runs parallel to the Via Roma, the main street, and is not a place for the squeamish. To get to the cobblestone street, one stumbles down uneven stone steps from the Via Roma. At night, those steps are dangerous, particularly in the blackout imposed under wartime regulations. This is especially true for the inebriated, who waver drunkenly on the steps, as they stagger from the bars back to their ships.

Our stay in Malta is short, about three weeks. Early in June 1943, we slip our moorings and join a convoy bound for Port Said. I am looking forward to seeing this city that guards the northern end of the Suez Canal, the gateway to the Far East.

The North African Coast

THE VOYAGE from Malta to Port Said is uneventful. The convoy heads east, past the island of Crete where we encounter long, slow swells off the coast. A tinge of seasickness spreads through my body and I'm glad when the island fades behind us. Now, only the mountain tips show over the distant horizon, and the swell has gone.

On the mainmast of the commodore ship flags flutter in the breeze as I watch from the boat deck. I'm off watch and lazing in the sun in a deck chair. The second mate's on the bridge above me. There's no mistaking his Geordie accent when he calls through the open bridge door to the helmsman, "We're changing course, Tommy. Follow the lead ship as she turns, lad."

The ship turns forty-five degrees to head southeast, and the sun is now on our port for'ard quarter. The ships, their bow waves breaking the calm surface of the sea, complete the manoeuvre smoothly and in uniform lines settle down to the new course. The Egyptian coast is roughly four hundred miles away. At eight knots, we should see the coastline in a couple of days.

Land ahead, low-lying, flat land on the distant horizon. A seaman points at it as he passes me. "Egypt, Sparks," he says. "We'll be in Port Said tonight."

The distance closes and the convoy alters course slightly as buildings and the minarets of temples take shape. Sunlight glints off windows and white-painted buildings. There is a beach, long and curving, to one side of the docks. The sand looks dirty from a distance and, on closer examination, there is also a seedy look to the crowded buildings along the shore front.

It is June 1943. The convoy heaves to outside the harbour. One by one the ships file into the harbour entrance. The wash of the bow wave spreads out and slaps against the long breakwater which juts into the sea. This breakwater, with the statue of Ferdinand de Lessops on the end facing the sea, is the entrance to the Suez Canal, the gateway to India and the Far East.

The sights, sounds and smells of Port Said both intrigue and repulse me. Everywhere there are jostling crowds of people, and merchants selling their wares and services in the streets and on the ships lying at anchor. The Ghili Ghili men, with their sleight of hand and conjuring tricks, produce eggs, coins and other paraphernalia from our bodies as they perform on the ship's deck.

Bumboats, small boats loaded with souvenirs, jewelry, bangles, and a vast variety of items, swarm around the anchored ships in the harbour. The *Empire Harmony* has its share as we lie at anchor. The boat people come aboard and sit on the decks with their wares arguing interminably over the price until one or the other gives up in disgust. "All right, Johnny. I give it to you. I like your face; last price."

The sale completed, the sad eyes of the trader light up as another customer is lured by the wares piled up on the deck. The lines on his wrinkled face deepen into a smile. It's an expressive face with a strong high-bridged nose, prominent cheekbones and forehead. On his head he wears a multi-coloured fez. Long loose clothing wraps round his body as he sits cross-legged on the deck behind his wares. "Hello, Johnny," he says with a calculating glance, "you want nice necklace for your girl?" He smiles as he barters, his grimy hand dangling the string of glittering glass beads in front of the victim.

Ashore in the crowded streets, Arab shopkeepers vie for our attention, and the inevitable voice calls, "Dirty postcards, dirty picture, Johnny? You like?" I ignore them, but they persist.

"No, Johnny? You no want? Look, look. Cheap, cheap, Johnny, I sell you cheap, how many you want?" The swarthy disreputable vendor fans out his pornographic photographs as he hurries beside us, intent on a sale, only giving up when he is convinced that we really don't want them, or is told bluntly to "fuck off."

On my second day in port, out of curiosity I buy some of the photos for a few piastres. The vendor has been following me as I walk alone back to the quay. His persistence pays off. We go down a side alley and he shows me the pictures. They are black and white and very explicit. "Okay," I say, "I'll buy them. How much?"

"Ten piastres, these, Johnny. You buy packet. I give you cheap. Fifty piastres."

We haggle. "Give you twenty-five."

"All right, You give me thirty. It yours."

I pay the money and pocket the packet of photos. Back in my cabin I look at them, and then grin. The top few are the ones he showed me, but the rest are just ordinary postcards. I've been had. Maybe there's a moral to this story.

In Port Said they seem to sell anything that moves. Young boys of eight or ten will even sell their sisters, or supposed sisters. They run up and tug at our clothing, peering up at us with big appealing eyes and earnest faces. "You want young girl, Johnny? You want fuck my sister?" Unabashed by a refusal, they continue, "Johnny, you want my brother? You give me money."

Our first visit to Port Said is not a long one, because we have only a partial cargo to load at this port. We will pick up the rest in Alexandria, about one hundred and fifty miles to the west on the other side of the Nile Delta. The transferring of cargo from the lighters brought alongside by harbour tugs goes on day and night. In less than a week we find ourselves heading out of the harbour mouth and along the Egyptian coast.

Alexandria has come and gone. It is now disappearing, a blur on the distant horizon behind the westward-bound ships. Fully loaded, the *Empire Harmony* is ploughing along through calm seas in a slow coastal convoy. Our destination is Tripoli, a port city, lying approximately twelve hundred miles to the west and about one hundred miles from the Tunisian border. At the convoy speed of eight knots we should reach the port in six or seven days. There, we will discharge our cargo and off-load some military personnel travelling to rejoin their units.

In Alexandria, I bought a small clay water container. Now it hangs on a hook under the bridge by our cabin. Semi-porous, it is exposed to the breeze set up by the ship's movements. I'm surprised how cool and fresh it keeps the water. I think the mate is envious. "Ah, Sparks," he says. Warning bells sound in my head, there's a gleam in his eyes. "That clay pot of yours, you know what it's made of?"

"Yes, clay," I reply, suppressing a chuckle.

The corners of his eyes crinkle and a grin spreads across his face. "It's formed out of clay, straw, and camel urine, then sun dried." He laughs as he turns away and climbs the bridge companion ladder. "Just think about it," he adds as he reaches the bridge.

The thought does not deter us. That jug with its clay plug is a prize possession. Ralph and I continue to use it for our drinking water.

The days pass, lazy and carefree under cloudless skies, as the small coastal convoy ploughs westward. The war has drifted north leaving the North African coast in peace. It is July and the sun smiles on us out of a deep blue sky. The sandy coastline, rugged in places with cliffs and gentle sloping yellow hills and sand dunes, lies constantly on our port beam. Occasionally, it disappears with the irregularities of the land mass and sweeps south as it reaches gulfs and bays; but like an old friend, it always returns.

It's the fifth day at sea, and I check the ship's noon position in the chartroom as I come off watch. We are averaging two hundred miles daily. The mate sees me poring over his charts. "Yes," he says, "We'll be in Tripoli tomorrow, Sparks."

On schedule, we arrive off Tripoli harbour. A deckhand swings a heavy sledgehammer at the large iron pin holding the anchor chain on the foc'sle head. It flies free, releasing the chain and hurling the two-ton anchor and its chain into the depths. The engine room telegraph rings for *full astern*. The anchor flukes hold, digging into the sandy bottom of the seabed; we lie at anchor, waiting outside the harbour for a berth.

We left the anchorage two days ago, and moved inside the harbour sea wall. It's a big harbour and we are anchored near the southwestern end. Our stern ropes are tied to the inner side of the sea wall. Number one and two holds are being unloaded into lighters. Fussy little harbour tugs tow the loaded lighters away while others are brought alongside to replace them. The Old Man and the mate are ashore today at a meeting with the military administration who run the harbour.

Word on our future movements has leaked out on the ship's grapevine. The *Empire Harmony* is to be stationed here in Tripoli. I heard the news from the steward after dinner. "The Old Man's talking about it now, Sparks. There's a conference with the deck officers in the captain's stateroom. I've just taken up sandwiches and drinks. He's going over procedures and duty rosters."

The mate confirms the rumour later that evening. He's with Jimmy when Ralph and I join them on the boat deck; these little gatherings are a habit we've got into when in port. "Yes," the mate says quietly, then pauses as he sips the beer in his glass. "Don't ask me how long. I haven't any idea, probably several months. We're starting our first job tomorrow. That's why the rush is on."

"Aye," Jimmy adds. "You can see for yourselves the damage done to the docks. The cranes and heavy lifting equipment have all been destroyed."

The deck chairs scrape on the wooden deck as Ralph and I pull them closer. It's pleasant sitting here in the evening outside the cabin with the stars overhead and a beer in my hand. After the cold wintry weather of the Atlantic and North Sea, the long summer months in the Mediterranean are ideal. "Think this port'll become a main base for Sicily, Chief?" I ask the mate.

"There's little doubt about that, Gord. There's Algiers and Alexandria, but we're the nearest big port so it should be busy as supplies build up."

I'm lost in thought a moment, because it's a question that's been worrying me for some time. Then making up my mind, I ask, "Yeah, I can see they need those jumbo derricks we've got, they're rated as eighty tons, but can they handle really heavy loads? I mean, things like tanks over eighty tons?"

The mate turns towards me, and I can see his eyes gleam in amusement as he draws on the cigarette, its glowing end momentarily lighting his face. "You wait and see. We're docking tomorrow. The dock's been cleared of debris and obstructions. Once we've tied up, the ship'll come alongside us and we'll unload her, tanks and all." He adds, "Those derricks working in tandem can load or offload the largest tanks. Ordinary ships' derricks can't touch them."

The following morning, the *Empire Harmony*, sandwiched between the cargo ship and the quay, unloads the cargo from the heavily laden ship. I watch from the aft deck rail as the giant jumbos work in unison. Heavy steel cables attached to the tank share the load between them. The steam winches take up the strain, and slowly the load rises off the ship's deck. Then, the jumbo derricks swing the load over our deck to deposit it gently on the quayside below.

A few days later, we are at anchor. The cargo ship has been unloaded and we've moved back to our normal anchorage inside the harbour sea wall. The mate's in the lifeboat tethered to the bottom of the gangway. Intrigued, I watch him. He is bending over the engine, tinkering with it. Unable to resist my curiosity, I walk down the slanting steps to talk to him. It is fatal, for as soon as he sees me his smile broadens and a cunning look creeps into his blue-grey eyes. He knows a sucker when he sees one.

"Ah," he says. "Just the man I'm looking for. I want volunteers to help the third mate run the liberty boat for shore leave. I'm making up a roster of names for manning the boat and a timetable giving the schedule of departures and arrivals at the main landing stage ashore." He pauses and looks up, a grin spreading over his face. "How would you like to help out?"

I jump at the chance, secretly delighted. With a look of satisfaction the mate writes my name down on his list. Ralph signs on too.

"You know how to operate the boat?" the mate asks.

"I've a motor bike at home," I answer with a grin, then add, "It's just a case of getting used to the gears. Clutch and gear shift are different. The throttle's a lever instead of on the handlebars. Yeah, can't see any problems."

He smiles. "Hmm, a motor bike, aye. What is it?"

"A 'bitsa,' BSA 500cc, bits of this and bits of that. Got a sidecar too. Hell of a thing on right hand bends. I prefer using it without the sidecar."

The mate snorts good naturedly. "Kids! All the bloody same. Had a bike myself when I was your age. Come on, we'll take the boat out for a trial run; you'll soon get used to it."

We practise awhile, using the controls and rudder as we race across the harbour then throttle down to slow the boat. The engine's an old side valve type but there's plenty of power in her. I learn to nose the boat into the dock, juggling the controls and rudder. Confident that I won't crack up his boat, the mate leaves me to it. I'm happy as I settle back in the seat behind the controls, ready to take on my first assignment. I've found a new toy.

A seaman sits in the bow with a painter and boathook near at hand, his hands gripping the sides. He grins, "Let her rip, Sparks." We laugh as I push open the throttle. The breeze is in our faces and ruffles our hair as we speed across the still waters of the harbour.

Nearing the quay, I throttle down and glide alongside the stone steps. The reversing propeller churns the water as I put her in reverse and touch the throttle to take off the way. It's a smooth landing. The seaman doesn't need the boathook.

He smiles and says, "Nice, Sparks, you're doing fine."

The words please me and boost my confidence. He steps ashore with the painter in one hand and ties the boat to a stanchion. My first passengers arrive and now I'm a part of the liberty boat crew.

We've passed her many times now while operating the liberty boat. She's a ship of mystery, an Italian passenger liner which was bombed and sunk in the centre of the harbour. She sits upright on the harbour bottom, the remnants of a once proud and beautiful ship. The gaunt, fire-blackened upper deck superstructure and masts are partly above water. Ralph and I have long wanted to explore her. Now, when we join Jimmy and the mate after dinner outside Jimmy's cabin, I decide to broach the subject.

The mate's in good spirits. He looks up as we approach. "Oh, no. Trouble's coming, Jimmy."

"Trouble? Us? Not on your life, Chief. It's that wreck over there." I gesture in the direction of the harbour as I sit down in a deck chair.

"Well, what's wrong with it?" he chuckles.

"Nothing, we just want to use the boat to explore it." I add hastily, "In off schedule times, of course."

The mate sighs, "Hmm, I thought you'd soon catch on to that. All right, but if I need that boat it's got to be here. Don't get the bloody thing holed on those wrecks. Be careful."

Ralph and I exchange glances and grin. The mate's already on another tack, something that he heard from another ship. The *Empire Harmony*, by virtue of its role as a crane ship, is literally an information centre which receives and dispenses news of merchant shipping operating in the Mediterranean.

The following morning, with Ralph acting as my crewman in the bow, we drop off our passengers at the landing stage on the quay. "Okay, Ralph," I call over the roar of the engine, "we've two hours before we have to return for them." Then, as I turn the helm and accelerate, I yell, "Hold on, we're off."

With our bow wave spreading across the harbour, we head for the blackened hulk. Throttling down the engine, I circle her slowly while Ralph takes up a position in the bow, the boat's painter in his right hand. He points to the ship's railings. "Port a bit, Gord," he yells. "We can tie up to the upper deck railings over there."

I push the tiller to starboard and watch the bow swing to port, slowing the engine at the same time. Then, judging the distance I take her in. When her bow gently nudges the railing, I briefly reverse the engine to take off the boat's way. Waiting his opportunity, Ralph jumps on the rusty charred deck and ties the boat to the rail. A smile of satisfaction flickers across my face as I switch off the engine and scramble aboard the wreck.

"Jesus, it stinks in here," I mutter.

Ralph wrinkles his nose, "Yes, it's rotting mouldy wood, or someone's used it as a latrine. Watch where you're stepping." We roam through the shattered cabins. The woodwork is burnt in places, splintered and gouged in others. Nothing but an empty lonely shell remains of the once luxurious upper decks, where passengers wined and dined in more peaceful times; the heat and fury of fire saw to that. Where the fire has not reached, scavengers have, stripping the ship of her former glory and leaving bare wrecked cabins behind. Graffiti covers some of the cabin walls, one to the effect that *Kilroy was here*. We are obviously not the first visitors.

The Backwater of War

LIBERTY BOAT DUTY has its ups and downs, depending on the alcoholic level of the passengers. The last run back to the ship in the early hours of the morning is the rowdiest. Drunken latecomers stagger back along the quay in small groups. Some are half-carried by their friends and dumped into the boat to be carted back to the ship.

Tonight is no exception. It's gone midnight and the sound of drunken voices carries on the night air. The liberty boat has landed another load of inebriated crew members. I can hear them stumbling and cursing as they make their erratic way up the gangway. The night is warm and my cabin door is open. Only the screen door is closed and a fan whirs in the background. Engrossed in a novel, I hear voices approaching but take no notice. Ralph's on duty tonight, and should be back soon.

The screen door slams open against the bulkhead. Startled, I stand up, dropping my book. The third mate, his khaki uniform soiled and shirt buttons torn off, staggers into the cabin, half falling over the raised waterguard of the doorway. Steadying himself he looks blearily around; he is in a boisterous, drunken mood. In his unsteady hands he carries a large flower pot which he has liberated from some night club. As he weaves about the cabin, the pot tilts, scattering earth on the floor. His breath stinks of beer. "Hi ya, Shparkshh," he says. "I've brought shomething for you. It's jusht the thing for your cabin."

Unable to keep the tone of revulsion from my voice, I say angrily, "Get that thing out of here. You're spilling bloody earth all over the carpet."

His mood changes and he becomes nasty. "Oh, you don't want the fucking plant, eh? I'll fucking smash your head in, you bloody snotty bastard."

Tired and angry, the pent-up anger within me comes out in a rush. "Fuck off yourself, you drunken sod. You're not worth dirtying my bloody hands on."

That starts it. The third mate flings down the flower pot, smashing it, then lurches towards me, his arms swinging. There are no Queensberry Rules in this fight. I know I have to get close in, or risk his kicking feet, so I grapple with him and we roll about on the floor trading blows.

I have him down at one stage, sitting on his chest pounding his dirty, foul-smelling face with my fists. An unreasoning anger possesses me; I want to smash him, to mark him. I don't like fighting, but now I'm involved I want to get in some hard blows. As I punch him, his hand thrusts up into my face. The fingers extend and he claws and gouges my eyes, breaking the skin around them. Blinded, I roll off him, the pain blinding me to further blows as the blood runs down my face. Shouts and voices cut through my reeling senses, and I feel hands grabbing me as crew members struggle to pull us apart.

Fortunately for me, the ruckus has alerted the crew and they break up the fight. As I am unable to see through my swollen eyes, the fight is becoming hopelessly uneven. The third mate is tough and muscular. If he had used his boots, I would have ended up in hospital, because that type of man fights to kill or maim.

The next morning when the third mate and I sit down for breakfast in the wardroom, the captain sees the cuts and gouges under my eyes. He grins, "Bloody hell, Sparks, what did you get hit with? Been ashore celebrating?"

He probably has a good idea of what happened, but I don't enlighten him. My only disappointment is that I have not marked the third mate as badly as he has marked me. The strange thing about the whole affair is that although I lost the fight, I have now gained acceptance by the crew. In their eyes, I have proved myself a man. The third mate is not liked among the crew, and several crew members approach me to voice their support.

The fight draws a new and unexpected ally to my side. It is, of all people, the second mate. After breakfast, I wander up on the bridge to nurse my wounds and hurt pride. The day is sunny and I sit in the Oerlikon gun pit on the starboard wing. Deep in thought with my eyes closed, I half doze in the warmth of the sun's rays.

The second mate, a short stocky man with cropped black hair, sits down beside me. I did not hear him approach and am startled when he speaks. "Daydreaming, Sparks?" he says.

His voice wakes me from my reverie. "Yeah, I'm afraid so. Didn't do so good last night. Got knocked about a bit."

"Oh, aye, I guess you're feeling sore!" he says. "I just want you to know I'm on your side, lad. I'm glad you stood up for yourself and didn't let the bastard put you down. You did the right thing; he's been asking for trouble ever since he came aboard." He snorts, "Huh, I never did like the son of a bitch. He's a nasty piece of work, and I'd love an excuse to plant him myself."

His words raise my flagging spirits and I feel a surge of confidence rising within me. I want to talk to this normally taciturn, reserved man, who previously seemed so remote and distant to me. "Well," I say, "you know, my only regret is that I didn't mark him like he marked me; I did my best, but he was too good for me." I pause, then go on. "I haven't fought anyone since I was a kid."

The second mate smiles, "You did just fine, Sparks. Kids fighting isn't so bad, but it's always different when men fight. The third is in his late twenties and an experienced street fighter. He fights dirty. Personally, if he gives me half a chance, I'll fix the bastard, but he knows that and keeps out of my way."

This is the start of an unusual friendship. The second mate is, in many ways, a forbidding-looking character, rough and ready, with a pockmarked, sunburnt skin. Brought up in the Tyneside, his deep-set eyes seem to pierce through you, but there is something about him that makes me instinctively trust him. I begin to feel a sense of belonging on the ship. It is a comforting experience.

On July 10, 1943, the invasion of Sicily begins. These are busy days for the *Empire Harmony* as she loads and unloads ships for the invasion force. Tripoli is a major supply base, but there are no air attacks. The Luftwaffe is too involved elsewhere, and the work goes on uninterrupted, day and night. With the fall of Messina to the Allied Forces on August 17, the Sicilian Campaign comes to an end and brings a brief lull to the port activities.

Life is beginning to get back to normal for the local civilian population of Tripoli. There is a military non-fraternization law, with "Out of Bounds" notices plastered on street walls in some areas of the city. The once-large Italian population keep to themselves, and the streets are relatively empty of civilians.

The subdued lighting casts a soft glow over the small dance floor of the night club. Chance has led us here, and a desire for change from the Officers' Club. Jimmy suggested it earlier this evening. "Let's go to a small night club that the mate and I found. It's nice and there's a band." Ralph and I readily agree. The mate's not with us; he's busy tonight on board ship.

The band plays and a spotlight is trained on a young Italian woman in a light blue dress, which sparkles as she moves. The song is "Lilli Marlene," a haunting melody that holds me in its grip. As she sings, the longings and sentimentality of the music and words play with my feelings. The song expresses the intense longing of my youth and the grief of war. When the song ends the audience applauds wildly.

I sit there absorbed. I am not alone in my feelings, for Jimmy is also affected. "My God, Jimmy," I murmur. "I've never heard that song before. It's fantastic, it gets right into your soul. It brings out something in me, the sense of loneliness, longing and yearning."

Jimmy nods, "Yes, I know. It does that to me, too. I heard it for the first time a few nights ago with the mate. That's also why I wanted you both to come. I thought you'd like it. It's a German song that's spread like wildfire amongst the troops here in North Africa. The Desert Rats claim to have captured it from the Afrika Korps."

With the end of the Sicilian invasion, there's a pause in activity on board the *Empire Harmony*, because the demands for a heavy lift ship have fallen off. The war has moved another step to the north, leaving us behind. The tedium of harbour life is broken when orders are received to proceed to Benghazi and Mersah Matruh with an urgently needed cargo. It's a short voyage, roughly five hundred miles to Benghazi and another four hundred farther east to Mersah Matruh. The *Empire Harmony* heads across the Gulf of Sidra. Three days later we sight the Libyan coastline and break off from the convoy to enter the port of Benghazi. Our stay is brief. The cargo is loaded into barges while we lie at anchor in the harbour.

The *Empire Harmony* sails at first light. The harbour breakwater slides past on our starboard bow and we're in the open sea. Once outside the port, the crane ship, along with other coastal vessels, forms up in convoy formation and heads east into the sun, through calm blue seas, under perfect August skies. It's strange to think that, only a short year ago, Rommel and the Afrika Korps controlled all this coast. Now it is quiet and peaceful, only the wrecked ports and damaged buildings, bullet and shell scarred by the street fighting and tank battles of bygone years, remain as evidence of the savage fighting over seemingly useless sand and desert.

The convoy runs parallel to the coastline from Benghazi to Mersah Matruh. We reach Mersah Matruh within the space of two or three days, and anchor off the small port to unload our cargo. This port, made famous for the desert battles fought around it, marks the turn-round point of our travels to the east. Within days the *Empire Harmony* is heading west.

By early September, we are back in Tripoli, back again to the more leisurely pace of this sleepy North African city. I'm glad because I like this lovely coastal town. I often roam its main streets and winding side roads, coming across old wreckage of burnt-out streetcars, and other debris of war. Remnants of the fierce street fighting are everywhere. Buildings still bear the scars of bullets, and piles of rubble lie where once fine houses stood. Outside the town, scattered in the desert scrub, are defensive positions and burnt-out military vehicles, tanks and half-tracks.

The late summer sun warms my skin through the thin khaki shirt. Ralph and Jimmy are on board but I'm happily exploring on my own, because I can take my time and satisfy my curiosity. Some of the buildings have little emblems cut into the stonework. They appear to be sheaves of wood with an axe head sticking out of the top. They are probably fascist motifs because most of them are damaged, purposely defaced, but their significance escapes me.

From a distance, I see a fire-blackened, crumpled heap of metal, lying on the side of the street. When I draw near I realize that it is a streetcar; the rails are twisted and its body rusted. It bears mute testimony to the savage street fighting. I look at the blackened, battered shell with growing curiosity and wonder what stories it can tell. Wandering on through the streets, my mind fantasizes on the recent past.

My brown shoes kick up the dust on the ill-maintained streets as I emerge onto a long sweeping boulevard which runs along the beach front. The wide sandy beach arcs around a bay with the Mediterranean, blue and inviting, as a backdrop. The beach is not crowded because it is a weekday. It's easy to tell the locals from the soldiers, even though they are all in bathing suits. Small groups of men dot the beach or swim in the water, while the local people stay in family groups and keep separate. I come here when I can because it's pleasant and I love swimming in the sea and sunbathing. On board ship, I live in a man's world and see the same men, day in and day out. It's nice to get out and away, and seeing young women at a distance is better than not seeing them at all.

A twinge of envy passes through me as I watch the antics of young couples diving off a floating raft, anchored just off the beach. I swim out to the raft and sit on it, dangling my feet in the sea. A lovely young Italian girl swims out to the raft followed by one of her family. She climbs on board, her black hair sparkling and sleeked back with water. Our eyes meet, we smile and I feel my heart leap. Then, she dives into the sea and disappears; the feeling is gone. I am left with an ache in my heart.

The deck seems to heave under me as I sit with my head in my hands. I feel awful. It's ten in the morning, the first day of 1944. *Oh, no!* I dash for the ship's railing and lean, retching, over the side. The captain slipped a mickey in my drink last night in the wardroom. This morning, the mate told me it was a mixture of whisky and gin.

The mate should know because I distinctly remember the captain telling him, "Aye, Mr. Mate, take young Sparks' drink away. He canna drink the New Year in with a beer; it wouldn't be right. I'll give him a real drink."

"Now," said the captain. "Here's to 1944. May the New Year see our safe return to the U.K."

As we drank the toast, I spluttered and choked. The drink was like fire water. My glass was still half full, a fact that did not escape the captain's eyes. He laughed, then said, "Come on, Sparks, you can finish it. Aye, lad, it's bad luck if you don't, you know."

I hastily drained the glass. As it settled in my stomach, the wardroom began to sway. I knew I had to get out. Laughter followed me as I headed for my cabin.

Stumbling along the passage and up the companionway to the upper deck, I reached my cabin door. Wrenching it open, I fell across the threshold, half in and half out of the cabin. I passed out and knew no more until this morning. My head throbs and aches. Someone had picked me up off the floor and put me in my bunk.

Rumours precede our departure, but it isn't until the captain returns from the Naval conference ashore that we know for certain that we are leaving Tripoli. It is late February 1944 when we finally hoist the anchor and sail into the unknown beyond the harbour breakwaters.

Once we are clear of the harbour, the captain breaks the seal on his orders in the chartroom adjoining the bridge wheelhouse. He reads them thoughtfully and turns to the mate. "Mr. Mate," he says, "we're routed to Augusta in Sicily for onward routing to Naples." Then he adds, "Tell the crew and also let them know there will be no shore leave in Augusta because we are joining another convoy there."

The news spreads rapidly. In the chartroom, the captain and the mate study the large-scale navigational charts and begin the task of laying out our route and course to Sicily, roughly three hundred miles to the north of Tripoli.

The familiar harbour and town of Tripoli drops away behind us as I watch from the bridge. The whitewashed walls of the town buildings and minarets become indistinct and finally disappear, merging into the North African coastline as the convoy heads north to Sicily.

A part of me is elated at leaving that small peaceful haven. At last we will be getting back into the war, which is important to me. I had felt the excitement of war passing me by and fretted constantly when I listened to the tales and exploits of other ships and seamen. Strange to say, I now feel a twinge of nostalgia. Leaving the port is like saying goodbye to an old friend. Tripoli was a bridge into another continent and another world.

Naples Bound

IT'S 2000 HOURS, and I've just come off watch. Tripoli has long since disappeared below the horizon. The seas are calm for February, and the convoy's making good time. We're ploughing northwards at a steady eight knots.

The radio cabin is just aft of the bridge on the *Empire Harmony*. I wander out and see the second mate in the chartroom. He glances up so I pause at the door to chat awhile. "Anything startling happening, Second?"

"No. It's pretty quiet, I guess the Jerries have too much on their plate in Italy to bother with us."

"Where are we now?" I ask, looking at the chart.

"Just passing Malta." The second mate traces the convoy's course with his finger. "There! Malta's a little to the west of us; we'll be changing course and heading northwest a few points soon. We should make Augusta late tomorrow afternoon, if all goes well."

We talk awhile before I make my way down to the late dinner sitting in the wardroom. Jimmy is now on watch, and Ralph's gone to our cabin. He'll be on watch at midnight, then I'll relieve him at 0400. It'll be awhile before we are again used to the watch system. After all, it's months since we were last at sea. Two or three engineers sit at the far end of the table talking quietly amongst themselves. They hardly look up when I sit down to eat my dinner, and they're still there when I leave for my cabin. Their cigarette smoke forms a haze above their heads while they relax over coffee.

Ralph is sound asleep in the top bunk, and entering the cabin quietly, I undress and slip into the bottom bunk. It's almost 2200; I'll be on watch again in six hours. Even the changing of the watch and Ralph getting up do not disturb me.

The convoy reaches Augusta late the following evening and anchors offshore. An anchor watch is set and the seagoing watches discontinued. There is no moon and in the light of the stars only the nearest vessels can be seen as dark shapes on the sea.

The next morning I wake to the sight of a low, flat coastline with a narrow coastal plain rising up into mountainous country. I look around in the early morning light. We are in a protected anchorage in company with other ships spaced out across the bay.

The breakfast table in the wardroom hums with talk. "Think we'll be here long, Jimmy?" Ralph asks as he helps himself to more cereal.

"No, the Old Man's going to a naval conference this morning. I reckon we'll be off tomorrow. It's the usual convoy procedures, call signs and frequencies, that sort of thing."

Jimmy must be psychic. Breakfast is no sooner over than a naval launch collects the captain for a conference. Armed with his briefcase he scrambles down a Jacob's ladder and steps aboard the launch. Later he returns with orders and convoy procedures for the final stage of the voyage to Naples. He grins at the crew watching him climb the rope ladder. "We're sailing at first light tomorrow," he says when he reaches the deck.

Sicily is quiet. From the anchorage, evidence of the fighting is not apparent, even through binoculars, but we are some distance from the shore and even farther from the towns and villages. The fighting zones have long gone, and are now concentrated on the Italian mainland, just north of Naples, where the Germans hold a defensive line across the calf of Italy.

We sail at daybreak. I am awakened by the sounds of the anchor chain as it is hauled up and locked into position. Another day has begun and I swing my legs over the side of the bunk, then head for the washbasin.

In columns of three abreast, the small convoy of ships hugs the Sicilian coast, heading for the Straits of Messina which divide Sicily from Italy. Close to the shore, on our port bow, the vast bulk of Mount Etna towers over the surrounding countryside; a heavy pall of smoke drifts from its cratered peak, and hangs over the land. I ask the mate if it's still erupting.

"No," he says. "But it's active, has been for a few months now. It has erupted a number of times throughout the last twenty years or so, long before the war."

Messina is on our port bow. The port and town loom large through the powerful binoculars I am using. I turn to the second mate, who has just taken over the watch from the mate. "Hell of a lot of damage ashore, Second. Both the town and port are really battered."

He looks through the binoculars a moment before answering. "Yeah. The Germans put up quite a fight here in July. It was the only escape route for them from Sicily." I nod as he adds, "The Strait's about two miles wide at this point."

The Messina Strait is roughly twenty miles long, and the convoy is through it in three hours. With the Sicilian coastline far astern, flags flutter from the masthead of the Commodore ship. "Humph," grunts the second mate to me, as he sees them. "Another change of course." He hurries into the wheelhouse, and I hear him pass on the new course to the helmsman.

Returning, he resumes his former position leaning against the bridge bulwarks. His body tenses as he watches the next ship ahead of us. It is altering course, and we're following in its path. We are on station; there's no need to reduce or increase engine revs. The second's satisfied and relaxes. He glances over at me. His pockmarked face breaks into a grin. "We should be in Naples this time tomorrow, if all goes well. It's roughly two hundred miles along the coast."

On schedule, Naples, with Mount Vesuvius in the background, comes into sight around midday. It's five days since we left Tripoli, and it is now early March, 1944, as we approach the Bay of Naples.

Beside me, Jimmy leans against the ship's rail. "Look, there's Vesuvius, Gord," he says. "That small island on the south side of the bay must be Capri."

We're still too distant to see detail, but the ancient cratered peak of Vesuvius and its grey ash slopes are clearly visible on the starboard for'ard quarter. The island, too, stands out against the background of the land mass.

With the rest of the ships in the convoy, we heave to in the assembly area just within the outer fringes of the bay. As the pilot boat approaches, a Jacob's ladder is thrown over the *Empire Harmony's* side. The engine reverses hard and the boat noses in to the rope ladder. A young American naval pilot waiting on its deck grabs the ladder and scrambles aboard. He grins at the small group of men, myself included, and waves when he crosses the deck and sprints up the companion ladder to the upper deck and bridge.

The distant ring of the ship's telegraph in the wheelhouse is followed by the answering ring of the engine room as they respond to the command. The engine note changes and we are on our way through the protective minefields that guard the port. The pilot launch has already left us and is now on its way to other vessels on its scheduled run.

On the bridge the pilot chats with the captain, identifying familiar landmarks. He issues commands to the helmsman about course changes when certain marker buoys come up on either side of the bow. With the methodical accuracy of long practice, he navigates us safely through the minefields and into the inner harbour itself.

The massive port facilities and sprawling slums and suburbs of Naples spread along the bay to the north. This is a major city crammed with shipping, and a main supply base supporting the Allied Forces, now fighting less than fifty miles to the north of us.

After the last vessel passes inside the harbour entry gates, a floating boom closes the channel. The boom has long steel wire netting hanging down from it, like an underwater curtain. It's designed to prevent U-boats and miniature submarines from entering the harbour, assuming they have safely navigated the minefields.

The mate is on the foc'sle head with the for'ard deck crew, while the second mate is aft with his men. Steam hisses from partly opened valves and gears clatter as the winch operators test their winches. The crew are ready to tie up the vessel, and stand by the coils of heavy sisal ropes and steel-stranded hawsers lying on the deck.

A flurry of orders from the bridge follows as the captain shouts over a megaphone, “Stand by fore and aft.” The ship, with tugs in attendance, gently edges alongside an overturned sunken ship lying on its side blocking the dock. The superstructure has been cut off the sunken ship, and a makeshift dock made with ramps running from the main dock onto its exposed upper side.

As we dock, I watch the activity around me from the boat deck. American uniforms are everywhere. Jeeps and army trucks with their characteristic white stars on the dull green bodywork, line the dock. I listen intently, because accents intrigue me. The unfamiliar harsh nasal sounds of American voices bark orders to the dockyard mooring crew as the heavy cables are fastened fore and aft to the temporary dock.

Early this morning, as soon as we had docked, the captain and mate went off to a conference with the military port authorities. They’ve been there all day and rumours and speculation are rife on board, but one thing is certain: the ship’s functions will be similar to those performed in Tripoli. It is late in the afternoon when they return.

Now, in the wardroom the mate confirms our assumptions. “Yes,” he says, “basically the harbour operations are similar to those employed in Tripoli, only we will be moored alongside this sunken vessel which the Port Authorities are using as a temporary dock.” He pauses, and adds, “Unlike Tripoli, we will not move from here, but incoming ships with heavy loads, such as tanks and other equipment, will come alongside us to be unloaded.”

He sips his drink, then says, “Any questions so far?”

One of the deck officers asks, “We under the Americans, Mr. Mate?”

“Yes,” the mate replies. “We’re under the U.S. Fifth Army. As we have been attached to them, we report directly to the Port Commander. I do not know how long we will be here, but, as the front lines are less than fifty miles away, I should think it will be at least several months.”

A murmur of voices rises as the mate finishes talking. After a few more questions the small gathering breaks up. Ralph and I wander out on deck. There is a nip in the air, but it is not cold, though I feel the need for a sweater. It is almost spring and winter is on the way out.

We stand by the rail on the boat deck outside our cabin. The outlines of the warehouses and ships at anchor in the harbour stand out in the starlight. There is no moon, and the blackout is strictly observed.

“Well, it’s not like Tripoli, Ralph,” I say, “We won’t be running the liberty boat by the looks of it.”

Ralph nods, "Yeah, that was fun." He pauses and adds pensively, "It's just about a year since we left the U.K. According to the mate we'll be here some time."

The sounds of laughter and loud talk come from the crew's quarters in the foc'sle head. The odd chink of light breaks the darkness as one of them slips outside on deck, opening and closing the blackout curtain covering the door to their quarters in the process.

Ralph's words awaken the feelings of homesickness that have been plaguing me of late, and I reply, "Yes, I would have liked to go home, but then, I suppose most of us would." I laugh to clear the lump in my throat. "Sounds like the crew are having a party tonight."

Ralph nods and turns away. "Yeah, it looks like it," he says, then adds as he heads for the cabin, "I think I'll turn in."

I stay awhile on deck, leaning against the rail listening to the sounds of laughter coming from the crew's quarters, and stare into the darkness shrouding the city and docks. My thoughts turn to Hazel. The shipping agent brought the ship's mail with him this morning, and now my mood changes and the depression lifts as I think of her.

I smile as I remember her letter and the photograph. She's grown into an attractive woman. It was a group photo of our families, but my eyes concentrated on her, sitting in front on the grass. Fair-haired and shapely, she was wearing a light summer dress which fitted snugly. I shake my head in wonder. *Uncle Will's daughter! We all used to play together. Uncle Will, not blood related, was Dad's old army buddy; they served together throughout the First World War. Hazel's just about family.*

A new thought strikes me as I turn and cross the deck to our cabin; I wonder if Dad ever got down to Naples? I make a mental note to ask Hazel when I write. Her father will know. Ralph's asleep, and my bunk light is on. He's got the top bunk and his curtain is closed. It's late, and after undressing, I roll into my bunk. I'll write in the morning.

Sleep just won't come; disjointed thoughts run through my head and I am restless. I ask myself, *What do I know about my father? Precious little.* I realize I hardly knew him. Then, a memory of him surfaces which belies the impression I have of an emotionless man. The day war was declared in September 1939 was the only time I ever saw him cry.

He sat at the dining room table, slumped forward, resting his head on his arms. Tears trickled down his face, wetting his work-stained hands. From our old battery radio set on the sideboard, the voice of Chamberlain, the Prime Minister, blared tinnily as he broadcast the declaration of war against Germany.

I was fourteen, and didn't understand. Alarmed and bewildered I tried to hug my father. "Don't cry, Dad, don't cry," I said, near to tears myself. I looked up at my mother as I felt her hand on my shoulder. "What's the matter, Mum? Why's Dad crying?"

There were tears in her eyes, too, as she gently pulled me away. "Come away, love," she said. "Leave him alone, son. It's the war, the bloody war, all over again. You'll understand one day."

Yeah, I think bitterly as I fall asleep, *I found that out soon enough.*

In the early hours of the morning, the undulating wail of the air raid sirens wakes us with a start. It's a far cry from Tripoli. There we had almost forgotten there was a war on. Gone is our haven of peace; we are back in the war. Jumping out of my bunk, I switch on my cabin light. Ralph is sitting up on the top bunk in his pyjamas as the sirens continue to wail. We look at each other questioningly.

"That sounds familiar, Ralph. Now what do we do? There's no point in manning the ship's machine gun posts; they haven't got the range. Anyway, the port's heavily defended by the army, and we're not a part of the defence system here."

"Yes," he agrees, "Nobody said anything about nearby air raid shelters either, so I assume we stay on board."

"Well," I reply, "That's about it. At least it's comfortable."

As if to contradict me, a series of deafening explosions make the ship shudder. "Christ, what the hell's that?" I yell, startled by the noise. Instinctively we run to the cabin door. There had been no whine of a bomb, but if it's close, the projectile reaches you before the sound does.

Opening the door, we see the bright orange flashes from Bofor anti-aircraft guns, firing in rapid sequence from the mobile battery located near our bow. The noise is ear-splitting and the blasts shake the ship.

We dress hurriedly and go outside to watch the barrage of anti-aircraft shells bursting in the night sky. Sheltering under the bridge overhang on the boat deck for protection, Ralph and I watch, fascinated by the scene. Pieces of shrapnel whine and buzz through the air, raising sparks when they strike metal.

It is an awesome display of firepower as literally thousands of rounds of cannon shells, mixed with tracers, scream into the night sky to meet the incoming enemy aircraft. They form cones of fire which concentrate on moving sections of the night sky, saturating each sector as the enemy aircraft move through. The shriek of falling bombs intermingles with the vicious crack and blast of anti-aircraft fire. Heavy explosions of advancing sticks of bombs light the night sky in sheets of flame and smoke when they hit the port and town.

Their bomb bays empty, the bombers turn and race for their bases. A second wave follows, and the pattern repeats itself. Then, an uncanny quiet settles over the port, as the anti-aircraft guns fall silent and the enemy aircraft withdraw. As if by mutual agreement the guns leave the sky to the night fighters who will seek and destroy the enemy. Almost every night, under the cover of darkness, the German bombers target the docks and town of Naples in an effort to disrupt the flow of war materiel and supplies to the Allied Forces in the front lines just north of the city.

We tune into the military radio channels where American radio operators report the location of enemy aircraft. The terms they use are unfamiliar and informal to my ears, but their meaning is clear. They refer to enemy aircraft as *bandits* and their operating procedures differ from those used on ships. Seated in front of the HF communications receiver in the radio cabin, I often tune into the army networks and listen to them talk in plain language.

The crackle of static is followed by the measured tones of the radio operator calling to *Charlie Dog* over the radio network, "Charlie Dog, this is Mike Echo. Do you read?"

Charlie Dog answers, "Roger. Loud and clear, Go ahead."

Mike Echo replies with a brief message, "Twelve bandits heading southeast, sector eight. Over."

The days turn into weeks and weeks into months, as the routine of hoisting tanks and other equipment continues throughout the day, and in an urgent situation, through the night as well. When there is an enemy air attack, the sudden wail of air raid sirens brings the work to a grinding halt with ships' crews and army personnel scrambling for cover.

It's the talk of the dinner table in the wardroom tonight. "Aye," says the second mate as he eats his dinner. "It happened just as we swung it over the side of the ship. The steel cable fastened to one side began to unwind and snap. 'Twas a bloody Sherman. That tank must weigh forty tons." He grunts and chews on a piece of meat, then continues. "The crew shouted at the driver in the cab as the cable strands began to break. He got out just in time before the tank fell off the holding cradle and flattened his truck."

I'd watched the incident from the aft midships walkway overlooking numbers three and four holds. It was as the second said. I remember the look on the face of the black American driver. He sat on a piece of broken stone pillar on the dockside, crouched forward, his elbows on his knees, and supporting his head with the palms of his hands. He stared at the flattened truck, dismay and disbelief written over his features.

A city under martial law tends to be dull. There is an abundance of regulations, curfews and non-fraternization rules. Naples is no exception in those months when we are there. I cannot recall going out in an evening to a movie or sitting in a cafe.

Merchant Navy crews seem to have been forgotten by the armed forces, although we are allowed to shop at the American PX stores. We can buy the basic items, such as soap, cigarettes, beer and other odds and ends very cheaply. Life, however, can hardly revolve round a PX store. Unlike Tripoli, we do not have the Officers' Club or its counterpart.

I have no doubt that some members of the crew find ways to amuse themselves in the evenings. There will always be some little nook in the harbour area where they can enjoy themselves at night in a pub or a brothel. The civilian population here is extremely poor and exists from hand to mouth; in these conditions, money or food can buy most things.

In the daylight hours, the town comes to life. From dawn until dusk the markets and streets are thronged with people. Bicycles clog the streets, and dilapidated country buses, roofs piled high with poultry cages, farm produce and bicycles, putter into town. The daytime also brings our world alive, and the war is momentarily forgotten. With the long summer daylight hours, shore leave beckons, even if only for a few short hours.

High above the ship's deck, the view of the Bay of Naples stretches out beyond the squalor of the docks and the harbour breakwater. As I work on the radio antenna, I pause after replacing a cracked insulator, and tighten the sagging wire. From here, I can see the island of Capri at the southern end of the bay where the headland juts out into the sea. There's a light summer breeze on my face and I brush the hair back out of my eyes. Then, shifting the safety belt around my hips, I readjust the metal clip on the ladder, and call to Ralph on the deck below. His face looks white and small as he glances up at my shout. "How's that, Ralph? Tight enough?"

His voice is distant, partly lost in the rumble of the winches as the deck crew unload the liberty ship alongside us. "It's okay," he gestures giving the thumbs up sign. I linger for a last look at Vesuvius; the towering ash-cone crater combines with the reflected blue of the sky from the surface of the bay. It lies to the south on the southern curve of the bay. The road to Capri and Sorrento winds past its long sloping sides. The urge to explore the countryside stirs within me as I start down the ladder.

Vesuvius and Pompeii

RALPH AND I are talking, so I don't hear Jimmy approach. "Hi," he says, nearly startling me out of my skin. He pokes his head round the cabin door and adds mischievously, "Oh, did I startle you?"

The soft Scottish brogue, though not pronounced, is unmistakable. I know instantly who it is. Our eyes meet as I turn round. "Startled?" I say with a laugh. "No, not really, just talking to Ralph. Come on in."

His smile broadens as he steps inside. "Sex, that's what it is. You young lads can't get your bloody thoughts off women."

"Christ," Ralph replies, "You're the one to talk, Jimmy. You and the women in your lodgings, just to mention a few."

Jimmy grins, then says, "I've news for you. The mate and I are planning a trip to Pompeii. Do you want to come?"

Delighted, I glance at Ralph. There's no doubt about his agreement. "Jesus, Jimmy, we'd love to, wouldn't we, Ralph?"

"'Course," Ralph replies, his face beaming like a Cheshire cat. "When're we going?"

"Friday morning, first thing," Jimmy says. "Aye, I thought you might like it. It's settled then. A company car will pick us up at the gangway."

It's Friday and the car's right on time. The mate and Jimmy have just joined Ralph and me. I feel like a kid on a Sunday picnic outing. In fact, Ralph and I were the first down for breakfast. The car's an Italian model, a black saloon, probably a Fiat. We move down the gangway, the mate in the lead, as it rolls to a halt at the bottom of the steps.

"Ah, Mr. Mate." The driver, a tall black-haired Italian, gets out and greets the mate as we file down the gangway. "*Buongiorno, Signor*, you have your friends. That is good."

The mate introduces him as Signor Luigi Marcioni, office manager of the shipping agency. He laughs and protests at the formality. "No, no! Call me Luigi; Marcioni too formal."

His lean face, sunburnt and weathered, puts him in the late thirties. It is his laugh and mannerisms which appeal to me; they are infectious and I warm to him. He and the mate are old friends, having met each other in the shipping office on many occasions over the last few months.

With a flourish, he ushers us into the car, and we're off to Pompeii. The mate is in front, while the rest of us are in the back seat. I'd almost forgotten how luxurious a private car can be after army trucks and jeeps. I catch fragments of speech as Luigi talks to the mate. His English is good, but his driving? Well, judging by the speed of his takeoff and apparent disregard for the rules of the road, it's not quite as good.

"Pompeii," he explains, "is only partly excavated. The war, she stopped all work on the place."

Through narrow back streets, past diversions where bomb damage blocks our route, we gradually leave the suburbs behind, and soon we are free of the sprawling city. In front of us are the flat, dusty coastal plains which lie between the shores of the bay and the nearby coastal mountain ranges. The road is unpaved, stony and dusty. A long plume of dust follows the car and hangs in the air behind us.

My eyes are riveted on the passing countryside. Cultivated fields edge the winding road, and vineyards dot the lower slopes of the mountains and plains. The car swerves violently to avoid a chicken, bringing my mind back to the road as I hold onto the seat in front of me. "Bloody stupid chicken," Luigi says angrily and looks at the mate. "They always try to commit suicide beneath the car. Why I try to avoid them, I do not know." He laughs, his humour restored. "Maybe I too soft hearted, is that how you say it?"

If nothing else, Luigi is dramatic, a born actor. He flings out his arm through the open side window. Pointing to the conical mountain just ahead, he says with pride, "There! Vesuvius. See the lava flows down its sides? Those flows came right down and into the sea in AD 79, two thousand years ago," he adds for emphasis. His dark brown eyes light up in pride at being able to show us the sights of his country.

The mate nods. "That's a long time ago. The soil must be very fertile around here, judging by all the vegetable and fruit stalls. Is that due to the volcanic ash covering the countryside?"

"Yes. It could be, but even in Roman times this land was very fertile. When the volcano erupted it not only buried the city and port of Pompeii, but it also buried Herculaneum, another town. Everything in the path of the lava and ash was gone, forest, vineyards, everything. It must have been a terrible sight at the time."

Turning off the main coastal road, the car jolts and rattles on a corrugated dirt track. We are heading into the slopes of Vesuvius. Luigi gestures ahead. "Pompeii is over there. It lies under the lava at the foot of the mountain. We will go to the right and park the car near the main gate. You know, this old town was here many centuries before the eruption. It is one of the best preserved towns of those times. Now, you will see for yourselves."

We draw up in a small cloud of dust. Luigi jumps out of the car to open the doors. A crowd of young boys gathers around us, clamouring to be our guides, but he chases them away with a stream of Italian. They hurriedly disappear, no doubt to seek easier pickings elsewhere.

Luigi speaks to the elderly ticket collector in rapid Italian. Then, with the tickets in hand, he beckons us through the entrance. We enter an old, old world in which we gaze upon the once mighty civilization of the Romans.

I stand in awe of the once magnificent communal bath houses. Massive in their construction and complex in their plumbing systems, they reveal a high technical skill. The lead piping leading from the cisterns and water catchment systems catches my eye. "Jimmy," I say in wonder, "I never knew they had lead piping in those days. Look at them, they're still intact after all this time."

Jimmy looks at the piping thoughtfully. "Yeah, but they also used lead piping for drinking fountains and household use. Many of them must have died of lead poisoning. That is, if other diseases didn't kill them first." He pauses, then adds, "Life expectancy in those days was only about thirty years."

Human forms, preserved in their death struggles by the encasing ash, are displayed in glass cabinets. My imagination runs riot in this place where naked men and women bathed in the long gone past. Even dogs and other animals are preserved. I shudder as I look at them, because the ruins have the musty smell of death. Luigi is standing near me as I look at a man in the show case. His body is contorted and the agony of death shows in every line on his face. "Luigi," I say, "how did they preserve these bodies in such detail? Look at the face, it's so graphic."

He nods, his face thoughtful. "Yes, it is," he says, then adds, "The archaeologists brought in experts who used the encrusting ash surrounding the bodies as a mold. Small holes were drilled through the mold and then filled with a quick drying liquid compound. Once the compound had dried solid, the mold was broken and they had a perfect image of the body's form and features at the time of death." He pauses, shaking his head in disbelief. "So simple and yet so effective." Luigi gestures with his arm, "Just look at them! They died two thousand years ago."

He half turns, including us all in his conversation. "Look over there," he exclaims. "See the agony in that woman's face. It shows the real terror which these people faced." Luigi pauses, then goes on. "It is the nature of things. Fortunately death was quick in coming because they stood no chance against the lava and choking ash. Some escaped by ship from the harbour, but most died in the earthquake and lava from the volcano."

Cobbled streets and the remains of stores, bakeries and all the implements of life line these streets, preserved in the places where they were found. The streets have sidewalks and at each intersection there are raised stone blocks with spaces between for chariot or cart wheels to pass. During the rains, pedestrians used these blocks as stepping stones to cross the street without getting their feet wet.

"Gordon!" It's the mate's voice. I'm several steps ahead of him walking past the remains of an old bakery. Turning, I see him pointing to a stone water catchment bowl on the wall of one of the shops. "What do you think that is?" he says.

I look at the bowl jutting from the wall and the stone carved opening above it. "It's a drinking fountain," I say. "There's plenty of them around. Why, anything special about it?"

The mate is patient, "Look where the water comes from."

Glancing up at the water spout, the light suddenly dawns, and I laugh, "Oh, all right, it's a woman's sexual parts."

Jimmy and the mate exchange a look of long lasting suffering. "He's seen it, Jimmy. He's seen it," the mate mutters in a teasing tone.

From the house of the two rich brothers, with its erotic mosaics, to the red light district, we wander through the streets. Here, phallus signs appear on street corners; placed on the walls of corner buildings, they point down the street. Luigi sees us looking at them with amusement and he laughs at our questioning glances.

"Ah, yes," he says, "Pompeii was a port. People from many other countries and many different languages came here. You can't mistake the meaning of the sign. It points the way to the red light district. It was very popular in Roman times with both the townspeople and visiting sailors." There's laughter in Luigi's dark brown eyes; they twinkle as he says, "Come, you'll like this, I show you."

Our curiosity aroused, we follow him through the streets. "Christ," says the mate as we see a gigantic stone penis, nearly waist high, mounted on the stone sidewalk and pointing down the street. "Look at that for size!"

Luigi stops and grins as we admire it. "This stone penis," he says, "shows where the brothel is." He points to a building about halfway down the street. "It is that large domed building. Let's go and see it."

Unconsciously, our footsteps quicken as we walk down the street. Jimmy and the mate are laughing over a private joke, and I am eager to see what these old brothels look like. It exceeds my imagination. We stand in the arched open entrance a moment, before moving down the short passage into a high domed circular lobby. Like the spokes of a wheel radiating from its hub, passages lead off from the lobby to small rooms. The mosaic murals above each passage catch my eyes. "Christ, Jimmy," I gasp. "Look at that!"

Jimmy laughs as I stand there gaping at the mosaic. "Aye, lad," he says. "It's only a woman being fucked in the ass. Look around you. There's another woman sucking a man's penis."

Luigi, laughter in his voice, says, "You paid your money and you took your choice. Is it not so today? But the mosaics, they are beautiful, aren't they."

We murmur agreement. The mosaics, remarkably well preserved, are works of art. Patterned in white and black pieces of marble, they depict the sexual activity and positions indulged in by the men who visited the women in the rooms leading off the passages.

Turning, Luigi leads the way down a passage. He stops at one of the small rooms. There are stone platforms set into the walls about two feet above the floor. He points to words carved into the wall. "Look," Luigi says, as he translates the words, "a seaman from Sicily has carved his name, occupation and home town in the wall. He's even included the date." He pauses and looks at us, smiling. "Human nature hasn't changed much has it? That's over two thousand years old."

I sit on the platform, and lightly touch the words with my fingertips. *That seaman must have sat just like this.* A shudder runs through me and I shift uncomfortably on the stone slab. *This must have been his and the woman's bed. I wonder what coverings they lay on?*

"Come on, Gord," calls the mate. "We're going."

"What do you think they put on these couches to lie on, Chief?" I say, rubbing my rear end. "It's bloody hard on the body."

"Rushes, something soft," he says with a chuckle. "Come on, get off your butt." I follow him out of the passage and back into the street. Jimmy and Ralph are ahead of us, walking with Luigi.

For fuck's sake. I'm nineteen now, I say to myself as I sit at the writing desk in my cabin. *I'm a man.* My elbows are on the desk and I'm holding my head between my hands. The letter is blurred. I can't write any more; my heart isn't in it. I sip the warm beer in my glass. It's half full, and I set it down carefully on the coaster.

It's early evening, July 1944. I'm dressed in khaki shirt and shorts, with open sandals. Summer is hot and sticky in Naples. No evening breeze stirs the air here like it does on the surrounding hills and high land. But it's not the heat and humidity that's depressing me. It is something within myself.

Laying my arms flat on the desk, I bury my head in my arms. *What's wrong with me?* I ask myself. *Why aren't I like other men?* There is a deep ache in my throat as depression settles over me again.

Is it really only a few hours since I saw her? I don't even know her name! But I remember her vividly. I'd gone to collect the ship's mail from the U.S. Army post office. Collecting the mail is something I like doing, because it gives me an excuse to stretch my legs and wander the streets of Naples.

It was in one of the main streets where I saw her, together with a girlfriend. As they approached, I noticed that she was wearing a light summer dress. There was something different about her, and I glanced up again as they drew near me in the crowded street. The sun glinted off her hair. Then it struck me. She was blonde. When our eyes met, I saw that they were blue. Most Italian women have black hair and brown eyes.

She was beautiful, and I felt my heart jump when she smiled. A momentary feeling of sunlight and warmth transformed the dusty street. Not wishing to appear rude by staring, I averted my eyes. Then, I heard a woman's voice and an arm linked mine as she greeted me with a laugh. "Hi," she said and her eyes smiled into mine, catching me off guard. "Would you like to go out with me?"

Flustered with the directness of the question, I panicked and stammered the lame excuse, "Oh, I, I'm sorry but I can't today. I've got to pick up the ship's mail." Why I did it, I don't know, but I pulled my arm free and walked quickly down the street. As I hurried away, I glanced back. She had rejoined her companion. I heard their laughter and my ears burned. I felt like an immature fool.

Back in my cabin, I can't get the scene out of my mind. I bite my lip and mutter to myself in frustration. *Why do I react like this?* I am not unaware of sex. It's the very opposite, but I seem to be frightened of women. My mind goes back to Juanita in La Linea. I was seventeen then, and the same feeling of panic had overcome me.

Where have things gone wrong? I think back desperately, trying to understand. *What makes Jimmy, and men like him, different? He thinks about sex, the same as I do, but he acts upon it, at least, judging by the stories he tells. He must live in a different world than I do.*

Yeah, maybe that's it. It's all the religious rot I've lived with, along with the moralizing platitudes of my mother against anything that smacks of illicit pleasure. "No, Gordon, you must not masturbate; it's wrong. It'll stunt your growth." She'd drummed that into me since early childhood. If nothing else, she succeeded in giving me a real guilt complex.

The beer tastes strong as I drain my glass and pour another. It's some stuff the ship's chandler got from the American PX store. My eyes are half closed, but my thoughts are active.

Where have the years gone? My birthday was three months ago. It's come and gone before I realized it. I didn't even get a birthday card. Who cares! I shift in my seat and turn my head in my arms. I feel tired of it all, and I'm bitter.

How long have I been at sea now? Christ, it's two years since my first ship. Almost three if I count my radio training and gunnery courses. What was that my mother said before I left home? Something about not going with women! Aaahh, yes. "Save yourself," she had said. "Save yourself for your future wife." What bloody wife? She'd carried on like a broken record. "You don't want to catch a disease, Gordon. You'll never have children, if you do." Mother made love sound like a dirty business, like it was for getting babies only.

Get real, Mother, I'm a normal healthy male, no different from anyone else. Except of course, that I'm cloistered in a world of men. Now, today, I had this opportunity and what did I do? I panicked and fled. Jesus Christ, I hate myself. In frustration, I hit the desk with my fist. My glass wobbles, spilling a little beer, and I grab it with my other hand. The action calms me, but I shake my head in wonder. *What a fool I am.*

My thoughts ramble on and inevitably I think of the nuns of St. Mary. *Those bitches whipped the hell out of me. Exorcising the devil, I think they call it.* I remember my mother saying that when I got angry I used to butt the nuns with my head. Perhaps that's why they tied my hands behind my back and beat me. *Something happened! Did they crush my spirit, and put the fear of God into me? Is it that fear of damnation that now holds me back? And then the Brothers of St. Egbert's College finished the job.*

I laugh behind the private tears that well in my eyes. *At least they can't control my thoughts and dreams.* The memories and nightmares drift through my mind, renewing my hatred. In my fantasies I am the one with the power; the tables are turned. Through this hatred I curse them for their power and authority that still exists in my mind. *I cannot rid myself of it. Why else would I react as I did today?*

Jimmy's voice breaks into my thoughts, and I look up. His dim form is outlined against the light filtering through the mesh of the screen door to the cabin. "You all right, Gord?" he calls anxiously as he sees me slumped across the desk in the light of a table lamp.

"Yes, Jimmy," I reply. "Just thinking. I've been trying to write a letter, but I can't concentrate."

"Come and join the mate and me. Ralph's with us."

Stretching, I get up and follow him across the intervening deck to join the little group sitting on the boat deck. My thoughts warm to Jimmy. He's a good guy, and so is the mate. They must think I'm a strange one, because I'm often on my own. There are times when I have to be alone; I need the space. I can't seem to get close to anyone. Perhaps I'm afraid to. I get hurt that way.

Twenty pounds? I shake my head as I read Mother's letter. The words stand out on the page. "Do you remember Lal Keyes?" she asks. "You know, one of my old campers before the war. Well, he and his wife visited us the other day. The poor lad had a tough time in France with the army, so I gave him twenty pounds." In a previous letter it had been Wamba, another of her pre-war campers. It seems to me the word is getting around that Mother is an easy touch, especially now that the second front has been opened up by D-Day, the invasion of Europe on June 6.

Rome was occupied a few days before the D-Day landings and now the fighting in Italy has moved farther north. From Mother's recent letters I gather she often buys rounds of drinks for her former campers up at the Owl Inn, the local pub. She's quite open about it in her letters, and even has the gall to write as if I am on a pleasure cruise, while her friends are fighting in France and Belgium. I don't blame her for having a drink or two with her friends, but it's not her money she's giving away. It's mine.

I'm worried and bitter, because that money was intended for food and other necessities, not for a bunch of spongers. My salary is the only source of income she has. I'm living on my danger pay of ten pounds per month so that my full salary can go into our joint account.

July has passed and now it's August in Naples. The hot Italian summer is at its height and there is a festive feeling in the air. Rumours are rife aboard ship. The current one is that we are to be relieved by a new crew shipped out from England. In late August, after the captain attended a meeting ashore in the harbourmaster's office, the mate confirms the rumours. "It's true," he said. "We are being relieved by another crew and will be signed off the ship's articles in a week or two." It is eighteen months since we left Westhartlepool in northern England, and it seems unreal to us, almost too good to be true.

On September 5, 1944, we sign off the ship and board a troopship in Naples harbour, bound for Gibraltar. The sight of the Rock brings back memories of the previous year, but we do not stay long. Transferred to another troopship, we sail for the U.K. The return journey is uneventful with only the occasional alerts. We have missed D-Day in Europe, but we were in the European war theatre long before D-Day, along with the Allied troops who were fighting their way up through Italy. Now the Germans are heavily engaged fighting on the Russian front and in Europe. They are too occupied elsewhere to worry about us. The troopship arrives in Liverpool in late September.

Jimmy, the mate, and Ralph have gone. The London train is waiting impatiently alongside the passenger platform, steam hissing from its old worn cylinders. I thrust my suitcases up the short flight of steps of the passenger coach and jump aboard. My heart is light and I'm eager to be on my way as the train gathers speed and heads south for London, the flying bombs and V2s. It's home.

Mediterranean & North African coasts

PART THREE

The Homecoming, 1944

IT IS MID-MORNING in early October. There is a wintry nip in the air, and I'm glad of my bridgecoat as I cross the open plains from Chingford and enter the forest. After eighteen months in the Mediterranean, I'm not used to the damp raw winters of England.

The ground is hard and frosty underfoot as I make my way along the familiar forest path. I stop to look at the tree I fell down when I was nine. Trying to save myself, I tore the skin off the little finger of my left hand as I slid.

I grin as I look at the scarred finger. The deformity has been noted on my Merchant Navy discharge book. In the *Tattoo or Other Distinguishing Marks* column are the words: "Scar on little finger of left hand." The shipping office official had laughed when he wrote in the sentence. "That's a good thing to have, son," he said. "If sunk and they ever find your body, that'll identify you." He had a sense of humour; black humour.

The old farmhouse still looks in good shape when I emerge from the forest into the small clearing. It's the weekend, and my brothers are playing with a football on the front lawn. They look round in surprise when I call out, "Hi, kids. Want to give me a hand?"

As I open the garden gate, they recognize me and come running, Peter in the lead, followed by Douglas, both yelling, "Mum, Gordon's here!" Laughing, I drop my suitcases and hug them, as my mother and sister Eileen hurry from the house.

It's during lunch that Mother tells me about the parachute mine. "Aye, love," she says, "the bloody thing landed close to the house during the night, and blew in the back door." She pauses momentarily to glance at the boys, "You two, eat your vegetables or you'll get no afters!"

Turning back towards me, she continues, "Oh, aye," she says, "apart from knocking tiles off the roof and the usual broken windows, there was no serious damage. But a few nights later, they dropped a cluster of incendiary bombs on us, which was worse. By the time the air raid warden and the patrol from the Bury got here, we'd put out the fire in the barn. I couldn't have done it without the boys." She looks at them, her features softening and tears gather in her eyes. "What with your dad gone, I just couldn't manage without them. If the barn had gone, so would the house."

I nod, for the barn is actually part of the house, forming one leg of an L-shaped building. Once a fire had got hold in the dry wood framework there would have been no way to save the house. The farm is very isolated, half a mile from the nearest road, with no water hydrant.

"Remember the oil bomb, Gordon? You know, the one that nearly hit the house?" Peter's voice is excited as he rummages through his treasure trove of war souvenirs. "This," he says, "is the nose cap. I dug it up out of the crater." His eager young face looks up at me as he holds the large metal cap in two hands.

Both boys are now in their early teens and growing up rapidly. Peter's fourteen, five years younger than I am. Douglas is eighteen months younger than Peter.

Now, taking the nose cap in my hand, I look at it. In the late summer of 1940, a German bomber laid a string of bombs which ended almost on top of the old house. "Yes, I remember it," I say softly. The vivid flash of the explosions and the fear gnawing at my belly, as I lay sprawled on the ground by the back fence, comes back to me. "It was near, that one. You and Doug were with Mum and Dad in the house."

Weighing several pounds, the nose cap is at least eight inches in diameter and has a large iron ring on the conical tip. Peter has cleaned it with a wire brush and painted it. It is the pride of his collection of shrapnel, incendiary bomb fins and spent bullet cases. He's so proud of his bits and pieces that I smile, because he reminds me of myself at that age.

I feel like an old man as my younger brothers clamour for my attention. Douglas, his short curly hair contrasting with the darker brown hair of his brother, has the tail fin of a small incendiary bomb in his hand, and thrusts it under my nose. "Look, Gordon, this one landed on the roof of the barn while you were away. Mum and Peter knocked it down from the gutter, and I helped to put the fire out in the barn."

"Yes," Peter's voice interrupts his brother. "Remember the night you went to the outhouse, and the Jerries dropped all those incendiary bombs? Dad reckoned they saw your flashlight."

"They must have!" I shrug my shoulders as I hold the remains of the incendiary, my fingers gently tracing its small curving tail fins. Then I continue in a soft murmured voice. "I didn't even get into the outhouse, but stood between it and the barn. The flash of the bomb bursting overhead lit up the sky and then I heard a strange humming whining sound, like a swarm of bees. Those small bombs were everywhere; there seemed to be hundreds of them." Pausing, my mind goes back to the lecture classes that I attended in the Air Cadets. "It was a Molotov cocktail. They're designed to burst about a thousand feet up and scatter hundreds of small incendiary bombs."

Peter nods his head gravely. "Yeah, Dad told me that, and he also said some of them blow up."

I look at Peter and feel a warming in my heart, he's learning, learning fast and no longer the little kid that used to bug his older brother. Mother's right.

They're good kids; they helped her save the house while I was away. They knew the risks.

"Yeah, that's true," I say. "When I got a shovel to try and put some of them out, Dad called me back into the barn. He said, 'Leave those bloody things alone; they are doing no harm out there. Some of them could explode.'"

I shake my head. "I don't know how those incendiaries missed the house and barn, there were so many. You must have seen them from the bedroom windows."

Peter laughs. "Yes, Mum wouldn't let us go outside, but we watched from the upper windows, until she caught us and made us go under the stairs."

"That I can believe. The back field and forest were dotted with the incendiaries; it was quite a sight. The anti-aircraft battery on the hill, together with other batteries, were putting up a barrage of anti-aircraft fire and shrapnel was falling. You could hear it hitting the roof and trees."

I hear Mother calling us. It's late in the afternoon, and she's made tea. I hand back the tail fin to Douglas and watch my two brothers scramble downstairs.

Yes, I think. *My family never knew the truth about that night, because I was too ashamed to tell them.* My face burns as I realize the danger I had put my family in. Driven by curiosity, I had signalled to the oncoming German bombers. As I crouched in the hidden space between the barn and the outhouse, I turned my flashlight on and off, making the letters B-O-M-B in Morse code, not once, but twice. Then it happened.

Fortunately, they only used incendiary bombs, possibly to see what was below, but if they had come in with high explosive bombs, my brothers and parents might not have survived. I should have known better. I was fifteen and had been in the Air Defence Cadet Corps for several years, so I knew about aircraft and explosives.

God save me from myself! I'm nineteen and still like a child blundering in a minefield. Right now I wish I were dead! I've stood Hazel up! The dreams I wrote her from Naples have disappeared along with the fantasies woven by my loneliness. It was so easy to put my thoughts and emotions down on paper, but now it's all different. Faced with my first real contact with her, I don't know how to act. Now I'm in a real mess.

It started the first day I was home. That evening I sat in the living room with my mother and sister. My two younger brothers had gone to bed. "I hear you've got yourself a girlfriend," said Mother as she poked the burning coals in the old Dover stove. "Young Hazel, Will's daughter. Bit old for you, isn't she?"

"Never thought of it," I said. "We're just friends. She wrote to me in Naples and sent me a photograph of you all on the front lawn. Anyway she's about your age, isn't she, Eileen?"

My sister nodded, her large brown eyes averted as she looked into the fire, and replied, "Yes, two or three years older than you."

"Twenty-two, if she's a day. Getting to be an old maid." Mother lapses into the Yorkshire brogue. "Look at your sister, she was married at eighteen and has a bairn upstairs. Oooh, aye, love, that one's after you. The lass told us all about the hot passionate letters you wrote from Naples. Moonlight nights and all the rest. She's after you, all right. Swanked about it, too."

I felt my face redden, secretly annoyed, and sensed hostility in my mother's voice. "No, you're reading too much into things, Mother. It's not that way at all."

"Well, maybe, but the Williams expect you tomorrow night for dinner," she said. "Will was your dad's army buddy and, as you know, the Williams are old family friends."

We went to bed late, but sleep did not come easily. My thoughts were too active and Mother's words churned in my mind. It was as if she was trying to crush any relationship between Hazel and me before it really started. I lay awake in bed, confused and disturbed because my letters were common knowledge, and now I had to face her parents.

I don't know why I asked Eileen to go with me to Hazel's home. Was it sheer panic, or fright? I don't really know. Meeting the Williams should have been no problem, but for some reason, it suddenly was.

I need not have worried. They welcomed me like a long lost son, and made me feel a part of their family. As the evening progressed, I felt my confidence returning. We talked long into the night about Italy and the Mediterranean, before Eileen and I left to catch the eleven-thirty train back to Chingford. Hazel must have thought it strange that I brought my sister, and, thinking back, I can't blame her.

Today is Saturday, September 30. I was supposed to go out with Hazel today, but everything has gone wrong. The day started uneventfully, until I mentioned that I was going to see Hazel, and that's when the trouble started.

"But I've told Grandma we're all coming to lunch! She's expecting us," Mother said. "It's your first weekend at home. Hazel will just have to wait; you can't disappoint your grandmother."

"But I promised Hazel, Mum," I said, dismay surging through my body. "Look, I don't want to disappoint Grandma, but I promised to take Hazel out. Why didn't you tell me? You knew last Wednesday that Hazel was expecting me this morning. I can't let her down, and they don't have a telephone." I looked at my watch. "It's eleven o'clock now, and it'll take me a couple of hours to get to Walthamstow."

Mother's face was set and determined. "We're not going to disappoint Grandma; she's old and she wants to see you, so that's that!" I started to protest, but Mother's not easy to stop. "Oh, don't worry," she said. "I'll tell Hazel what happened. They're old friends, and they'll understand."

Hesitant and uncertain what to do, I fell silent, but my instincts told me otherwise. I shook my head and gave up, leaving it to Mother.

Now, ashamed and humiliated, I am not in a talking mood and Mother knows it. I need time and space to think; the hurt still runs deep. *Mother's treated me like a child; perhaps I am a child dressed as a man*, I think bitterly.

It's not really cold enough for a fire but I've lit one anyway. Chopping the wood has expended some of my anger. Now the flames are growing as I feed the fire. October will soon be here, and there is a chill in the air. I wish I were back at sea, in the world I know and understand.

My mind drifts back to the scene in the woods this afternoon on our way to Grandma's. We'd just about reached the bridge over the stream when I saw Hazel coming towards us. My heart sank; I knew I was in trouble and deservedly so. The full import of what I had done fell on my shoulders. Eileen, along with her young daughter and my brothers, moved off up the path leaving my mother and me with Hazel.

Almost crying, Hazel said, "I thought something was wrong the other night when you brought Eileen with you. But I never thought you'd do something like this!"

Then, as she burst into tears, Mother consoled her. "You're better without him, love. Come on, we'll go back to the house, just you and me." Mother glanced at me. "You go with Eileen and the children to Grandma's. I'll see you there later."

A knock on the door interrupts my thoughts, and Mother pokes her head round the door. "I . . . I wanted to say I'm sorry about today, love," she says. "I didn't mean it to turn out like this. I just didn't want you to disappoint Grandma Mumford; she wanted so badly to see you."

I shrug my shoulders. "If you had told me earlier, I wouldn't have arranged to meet Hazel today. I've not been home even a week and you're already trying to manage my life." I look up. "You've got to realize I'm no longer a child, I'm nineteen and grown up. If I am man enough to fight for my country, and be the sole breadwinner, then I'm entitled to be treated like a man." I pause, my anger subsiding.

Suddenly tired of it all, I say, "Oh, what's the use, it's done now. I shouldn't have let you talk me into it in the first place. I should have told you to go to hell. I'd no right to stand Hazel up. Poor kid."

"Aye, Gordon," Mother says, "I shouldn't have done it, but I'm a mother, and you're all my kids. Let's not fight; I'll not do it again."

We talk on awhile, and then she leaves. I watch the door close behind her, and shake my head. I am beginning to realize how devious and scheming my mother can be. The sparks fly as I poke the fire and recline in the armchair. Restless, I can't settle down. Mother's back at her old game of divide and rule, and I'm just beginning to understand its implications.

"Yeah," Tony's face is thoughtful as he toys with his drink on the bar counter of the Palladium Theatre. "Remember Sid? He used to work under your dad at Alan's Nursery. I wonder what happened to him?"

I shake my head as I reply, "I don't know. I lost track of him after Dad died." Pausing momentarily, my fingers subconsciously twist the glass in my hand as I watch the beer suds swirl. "So much has happened over the last two or three years." I laugh bitterly. "Now, either my friends are dead or they're scattered throughout the bloody world."

Tony's about my age, nineteen, maybe twenty, with an angular, attractive face and dark hair. Slim and fit, he looks smart in his uniform. He's a naval lieutenant now, and, like myself, was just a young kid working part time in Alan's Nursery when I met him.

"I guess it's just chance, Tony, when you come across old friends," I say as I glance up at him. "Like you and I meeting on leave at old Alan's place. I just went down there to see how he was getting along. I like Alan and his wife and the kids, Barbara and John. Then, of course, you were there." The intermission bell summoning the patrons back into the theatre breaks into our conversation, and we join the surging crowd. Tommy Trinder and his chorus line of young, scantily clad women are on stage.

It's three weeks since I came home, and now Tony's gone back off leave to his ship. His companionship and our common interests sparked new life into me as we toured the theatres of London, and enjoyed the brief time we were together. With Tony gone, I'm on my own again and I feel lost.

By custom in the rural areas of England, the postman delivers both the mail and newspapers to outlying farms. Today is no different, and the postman has no sooner left the house than Mother opens the newspaper and settles into her favourite chair by the kitchen table. I am sitting opposite her at the table, reading the comic section, when I hear her cry out.

"Look, look at those poor women," Mother screams, flinging the newspaper to the floor. She bursts into tears. "The bloody mobs in Paris are stripping women and shaving their heads." Sobbing, she says haltingly, "Why? Why? For being so-called collaborators? Why don't the police or army stop them?" She buries her head in her arms on the dining room table as she slumps in her seat.

Aghast, I pick up the *Daily Express* and stare at the photograph on the front page. Stripped naked, their heads shaved, a line of women sit on a bench, watched by a jeering crowd. They are guarded by gendarmes who stand there, posing as if guarding cattle. Shame and revulsion surge through me invoking anger and loathing. *Why do the Allies and our own troops allow such things to happen? They have the power and arms to control these mobs, but don't. How can our troops stand aside and ignore these horrors?*

Now, as I look at the pictures and again read the captions beneath them, I realize that these photographs are old ones. They had been taken when Paris fell to the Allied Forces on August 25, 1944; however, they have only recently been released for publication in the British newspapers. I had heard that these things had been happening all over Europe, but these photographs have brought back the brutality and horror of war.

The realization makes me feel sick. I cross the room and put my arms around Mother to comfort her. I know how deeply she feels for she, too, has no hair, no eyebrows and no eyelashes. Mother had lost all her hair when my youngest brother, Douglas, had been born back in the early thirties. It never grew back after that. That's why she wears an old leather pilot's helmet, although she has a proper wig to dress up in.

"Don't cry, Mum, don't cry," I say, on the verge of tears myself. "I . . . I don't understand it either. I've never seen this side of war before. The war at sea is fought between men, not against women and children. The sea buries our dead and dying. The horrors we see are the wounded and dying on board, or the bodies floating in the water."

"Yes, I know," Mother says, "but how can men do that to their own people, and how can our people stand by and do nothing to stop it?"

"I don't know, Mum. I just don't know." The shuddering in Mother's body subsides and she gets up wearily from her chair. She looks at me, her face stained with tears. "I'll be all right. I just don't understand this world."

As I watch her leave the room, my mind goes back to a scene which happened not long ago in Regent Street, the heart of London's West End. Mother usually wears the pilot's helmet, but we'd gone up to central London together for the day sightseeing in the old City of London. It had been a long time since she had been in the City, and she'd dressed up for the occasion in a long beige coat, gloves and walking shoes. She wore a matching beret and had used a long ornamental pin to attach it to her wig. I wore my uniform that day because I knew it would please her.

The streets were crowded with shoppers and office workers. Mother's arm was through mine as we walked along the street, window shopping. Then, the unthinkable happened: a sudden gust of wind blew off her hat and the wig with it, exposing her bald head. She cried out and tried to grab her hat when it flew off, but it skipped down the street with Mother after it. Startled, I joined the pursuit, but she shouted at me, "Go back, Gordon, go back. Leave me; don't look at me. Pretend you don't know me." But I ignored her plea and tried to help.

Retrieving her hat and wig from the gutter, she jammed it hurriedly on her head. I'll never forget the frightened, horror-struck face of a child clinging to her mother nearby. Putting my arm around my mother, I helped her up. "Come on, Mum," I said, "we'll find a cafe. There's a Lyons not far from here. You'll be able to tidy up there, and a cup of tea will help you to feel better."

My thoughts return to the present, and I shake my head. I just can't understand.

London has not changed much. Since the invasion of Europe, flying bombs, commonly called doodlebugs, have replaced raiding enemy aircraft. Powered by ramjet engines, these pilotless winged bombs streak across the night sky at four hundred mph.

Early in my leave, I saw my first doodlebug. The air raid siren had gone, but the suburb was quiet until I heard the distant harsh noise of its engine. Then I saw the doodlebug with its long jet flame flaring out behind. I froze when I heard the engine cut out; then, mobilized into action by the distant whistle of air through its short stubby wings, I took cover behind a drainage culvert.

The bomb smashed into some houses a few streets behind me and exploded in a ball of flame. I got to my feet shakily, and heard the distant shouts of air raid wardens, followed by the sound of fire engines and ambulances speeding through the streets to the site of the explosion. I hesitated a moment, undecided, and then continued on my way. The rescue teams would be there before I could help.

There is no defence against the V2s. When a rocket with its built-in guidance system is fired from a base behind enemy lines, there is no warning. There is also no time to take cover. You are alive, or dead. Fortunately as the German lines are getting pushed back across Belgium and France, these rockets are now fewer and their effect is limited.

It's early November, and I feel impatient and unsettled. Life in the farmhouse has settled into its normal routine and the occasional visits to London's West End and its theatres are beginning to pall. It is depressing when there is no one to talk to or share the excitement of the city. My friends are either away in the services or, like John Cummings, dead. I am looking forward to going back to sea. It is in my blood now, and I need the companionship of the world I know.

I think Mother recognizes the restlessness in me and smiles when the cable arrives in mid-November. "It's an oceangoing cargo ship, the *Empire Path*, Mum," I say. "I'm joining her on November 22nd. She's berthed alongside Tilbury docks."

"Aye, son," she says, "I guess it had to come." The smile is no longer on her face; she looks weary. She pauses, then continues, "Maybe you won't be away so long this time, Gordon."

I nod. "Well, as you know, they never tell us until we are at sea. It could be anywhere; it is not a coaster like the old *Soborg*, which is a relief." I smile, "Maybe we'll be going over to the States, or to France or Belgium." I grin and shrug my shoulders as I see the worried expression on her face. "Don't worry, Mum, I'll be okay."

Wednesday, November 22, is another cold wintry day. The sky is overcast and traces of fog linger in the forest as Mother and I pick our way along the path, taking care to avoid the soft, spongy patches. The bridge over the stream echoes with our steps, breaking the quiet of the forest, then all is silent again as we pass by. *For me the bridge has come to symbolize my father, and unconsciously I say goodbye to him. I wonder if my mother feels his presence; I'm sure she feels it in other ways, because she has changed since his death.*

We reach the road at Sewardstonebury, a small village half a mile from the farmhouse, and I say goodbye to Mother here because it is another two miles to the station. She waves and stands still, watching me go with my suitcase in one hand and a carryall bag slung over my shoulder. The chill in the air nips my ears as I settle my uniform cap firmly on my head. When I reach the bend in the road I pause momentarily to look back. I see her wave and turn, disappearing into the forest. She is wearing the old leather helmet and a thick navy blue windbreaker.

Alone with my thoughts, I turn and set off down the road to Chingford and the railroad station. Ahead are a new ship and shipmates. I step out, my stride lengthening. I am nineteen and eager to be off. A surge of excitement fills me as I hurry down the road. I should be in Tilbury and on the ship by midday.

En Route to the Scheldt

"CHRIST ALMIGHTY, slacken t' bloody hawser, man!" Startled out of a deep sleep, I sit up in my bunk. The bosun's on the boat deck just outside my cabin, and I recognize his strong Gaelic brogue as he yells to his crew on the deck below. I'm halfway out of my bunk, still groggy with sleep, before the realization hits me. The ship is scheduled to leave the Tilbury docks on the first tide, and the crew are preparing for departure.

The vessel lurches. Now, fully awake, curiosity drives me aloft. I don heavy winter clothing and hurry from my cabin to the boat deck. It is still dark, but I can see the fussy little tugs. Their red and green eyes glow in the darkness, and their towing hawsers are secured fore and aft to the ship's bollards. The crew, with the mate in command on the foc'sle head, have cast off the lines holding the ship to the dockside.

The hawsers and ropes twang and stretch with the strain as the tugs pry the ship from the loading berth. Black inky water stretches in the widening gap between ship and dock. The bridge telegraph rings as the pilot prepares to manoeuvre the ship under its own power.

I think of Mike, and wonder if he was up all night supervising the raising of steam in the ship's boilers. He is a junior engineer, and we have become firm friends since I joined the *Empire Path* a week ago as the second sparks. He's about my age, as is John, the third sparks aboard.

A blast from the ship's siren echoes across the docks; then, free of the tugs' hawsers, the *Empire Path* edges out into the river as the first grey light of dawn chases the night from the skies.

It is November 29, 1944. The ship, fully loaded to the plimsoll line with military equipment and high explosives, heads down the broad muddy waters of the Thames to the open sea. My heart beats faster as a combination of excitement and underlying apprehension stirs deep within me. Perhaps it is the mysterious shadowy shapes of the docks as they pass, shrouded in the thinning fog, which cause the tightening in my stomach. The edge of fear only adds to the adventure of sailing into the unknown and a sense of satisfaction pervades my body.

My imagination is fired. Although we know nothing of our destination, except that it is somewhere in Europe, all is right with the world. The omens are good. Have I not met Mike and John? We are already a threesome, and, in the short week before sailing, the three of us haunted the dance halls of New Cross and the theatres of London. I've even met Karen, Mike's sister. Fair-haired and attractive, she is a year or two older than Mike, and looks smart in her army uniform.

The dawn river mists are beginning to clear as the cold watery sun pierces the overcast sky. A stiff breeze is blowing from the north. The third mate leans against the bridge bulwark, the hood of his duffel coat folded down. The wind ruffles his sandy-coloured hair as he stares over the protective windbreak on the bridge.

War-hardened and experienced, he began his sea career as a deck apprentice several years before the war. I think of him as a much older man, but he is only twenty-eight. The wooden buckles of his duffel coat are fastened, and his spare lean frame is hidden beneath the folds of his winter clothing. In the strange jargon of the seas, he is known as the third, but I get to know him as Steve, Steve Mathews.

Like Steve, I, too, am prepared for the weather and huddle in my bridgecoat against the wind as I look over the weather shield, watching the docks disappearing in the thinning mists. Oerlikon gun nests share the wings of the bridge with us, their barrels pointed menacingly at right angles to the ship. The faint clang of the engine room telegraph sounds through the sheltered wheelhouse door, and we can hear the muffled voice of the pilot giving instructions to the helmsman, "Two points starboard. Steady as she goes."

Jim, the chief sparks, is on the eight to twelve watch. I saw his long legs and thick-set form at the radio operating desk when I passed on my way to the bridge. The radio cabin is next to mine, on the bridge deck of the ship just aft of the wheelhouse and chartroom.

The third mate looks for'ard and nods towards the liberty ship ahead of us. "She's fully laden," he mutters as he glances towards me. "Never did like sailing on those ships: they're welded. The Kaiser Shipyards in New York churn them out one a day. The bloody bitches break their backs in a good storm. You sailed in them, Sparks?"

"No, but I've heard about them. There must be hundreds around."

The third laughs, a brief cynical laugh, then says, "If they make one voyage across the Atlantic before breaking up, they are considered to have served their purpose." He adds bitterly, "It doesn't matter about the crew; they were and are expendable."

Lost in thought, he watches the ship ahead, his eyes taking in the churning wake of its propeller buried deep in the water. Then his features soften and he smiles before continuing, "We don't have those problems with this ship. Its hull is half-riveted, half-welded. She's strong, unlike that vessel! I've sailed in her across the Atlantic in some of the worst weather, and she takes it in her stride." His voice reflects his pride in the ship as he adds, "She was built in Canada."

The river broadens as we approach the Thames Estuary. The low, flat marshy lands of Essex on the port bow and Kent on the starboard recede in the distance, as we pass Canvey Island. I pick up the bridge binoculars and focus on the shoreline. The flat-faced white buildings and red roofs of the seaside town of

Southend come close. I study them a moment, trying to pick out familiar buildings and the beach, but the distance is too great. The small coastal town brings back memories of my childhood, and with a sigh I hand the binoculars to the third mate.

"I used to play on that beach before the war, Steve," I say. "My parents would bring the family down to the coast in the summer. It seems a long time ago now. There's a little town called Lee-on-Sea, just west of Southend. I played cricket there for my school every year."

Training the binoculars on the shoreline, Steve scans the beaches. "The only time I went ashore in Southend," he mutters, "was for a gunnery course back in 1939."

A memory of mown lawns and the sand bunkers of a golf course taken over by the military as a gunnery school, flickers through my mind. I laugh as I turn and glance at Steve. "They're still training Merchant Navy crews there. At least they were in 'forty-two," I reply. "I took a two-week course on machine guns on the naval base, and in the dome near Tower Bridge."

A flurry of activity interrupts our conversation. We have arrived at the convoy rendezvous point in the estuary. The third mate hurries into the wheelhouse to join the captain and the pilot. The ship's telegraph rings *half speed ahead* and I hear the engine room answer as the pilot prepares to leave.

The sky is overcast and the sea is beginning to pick up. I feel the movement of the ship in the ground swell off the coast, and then watch the pilot as he scrambles down the Jacob's ladder slung over the side of the ship. He dangles there three or four feet above the waves while the pilot boat nudges its bow alongside the ship. Choosing his moment, he jumps nimbly onto the deck.

Turning back from the ship's rail, I watch the distant land falling farther astern. *Yes*, I think to myself, *it's a long time since I played cricket in Lee-on-Sea.* I remember with pride the day I bowled five batsmen out in one over. I was a hero that day. A warm feeling tinged with nostalgia fills my body.

I am nineteen now with three ships to my credit. This is my fourth ship. Despite the inner tensions and fear that lurk in my subconscious, I look forward to the prospects of enemy action. It makes me feel a part of the war, and gives me something to share with my compatriots. The enormity of dying never really occurs to me. Not that I have not been near death. That time on the tanker I felt its breath, but it was a moment of passing terror.

The area is quiet, with no enemy activity to confuse the ships as they manoeuvre into convoy formation. The Luftwaffe are busy elsewhere, and, because the E-boats have lost their bases on the French and Belgian coasts, they no longer menace the eastern shoreline of Britain.

The *Empire Path* is in position awaiting the commodore's command to sail. Shortly after his return from the last-minute naval briefing ashore, the captain breaks the seal on his orders. For security purposes, our destination is not revealed until the ships have put to sea. After all, *the walls have ears*, or so the posters say. We are bound for Antwerp, via the Scheldt Estuary in southern Holland.

Formed in several columns, the convoy steams steadily through moderate seas on a northeasterly course to Holland and the fluid front line, less than two hundred miles to the north. The Germans still hold a large part of Holland and the northern part of Walcheren Island, but the Scheldt itself is now clear of enemy forces.

The captain is a burly man in his mid-forties. I find him a bit intimidating as he glances around the table. His north country accent cuts through the low hum of conversation, and commands our attention. "The first seagoing convoy to dock in Antwerp just reached there yesterday," he says. Then he adds, "This will be the second convoy into Antwerp. There may be trouble, but on the other hand we may be lucky. We do have air cover, at least in the daylight hours."

As he turns and talks quietly to the mate, conversation picks up again. I lean back in my chair, half turning to John, the third sparks, sitting beside me. "Looks like trouble ahead, John."

He nods, "Yes, I guess we can expect that; it's pretty close to the front lines."

As we eat, we talk, mainly about the girls we met at the dance in New Cross just before we left. Mike, the junior engineer, was with us. "Did you take her home, Gord? You know, the one who dances too close!"

"Oh, you mean Julia. Well, only to the station. Funny thing that, a policeman moved us on when we stopped in the doorway of a shop in the High Street."

John laughs, then says, "She seemed a bit of a teaser, the way she pressed into you."

"Yeah, I think you could say that. She asked me if I'd like to go home with her. She has an uncle up in the city, somewhere near the Elephant and Castle. But no, I didn't go. Maybe I should've. I wonder how Mike got on with his girl?"

I liked John from the moment we met in Jim's cabin. Lively and humorous with an attractive face, he's about my height, with wavy brown hair and grey eyes. His face is animated and excited as we talk. He's like Mike in many ways. Both of them are full of life and fun to be with.

Looking back, I feel I've been fortunate in more ways than one. I passed my Second Class Marine Radio examinations while on leave, and now have my peacetime certification. I owe a lot to Jimmy, my last chief. He encouraged and helped Ralph and me with our studies.

I'm on my way to the bridge and pause on the quarter deck for a moment. The seas are roughening, and I watch the ship's bow cut cleanly through the oncoming waves. The sight of the surrounding ships and escort is comforting.

In the chartroom the third mate is studying a chart on the table. He glances up as I enter the door, and beckons me over. His finger traces our course across the large scaled chart. "Look, Sparks," he says, "this should interest you. There's the Scheldt Estuary. We'll be off the Dutch coast early tomorrow morning."

My eyes follow the pencilled line. "Where will we pick up the pilot, Steve?"

He points to a cross marked in the mouth of the estuary, "Right there."

"Hmmm, doesn't look like there's much room to manoeuvre if we're attacked either in the Scheldt or in the river."

Steve gives a short grunt. "Huh! No, there's no way to escape; we'll just have to sit there and take it, like a line of bloody ducks. But, like the Old Man said, we do have air cover during the day, and hopefully naval minesweepers are busy clearing mines."

He thinks a moment, his elbows on the table and chin in the palm of his hand, then continues, "We'll be on maximum alert going through the Scheldt and down the river into Antwerp."

The geographical outlines of the navigational chart show every detail of the surrounding countryside and shoreline. Walcheren Island with the small port of Flushing on its southern shore form one part of the northern bank of the Scheldt. The chart stares up at me, enhanced in the glare of the adjustable desk lamp clamped to the table. Encouraged by his words, I ask, "What about enemy naval forces on the Dutch mainland? I've heard there's a midget submarine and E-boat base on Walcheren. Do you know roughly where it is?"

He studies the chart for a moment, then circles an area on the northern coast of the island. "I don't really know, but it's probably in this area, near West Kappel." He shakes his head in indecision, then continues, "They can also operate out of Rotterdam and other nearby ports. Personally, I think our main problem will be mines in the Scheldt and air attacks." He looks up from the chart and grins. "I guess we'll have to wait and see when we get there."

Governed by ship's bells, the familiar routine of shipboard life returns in its comforting monotony. The regulated confining spaces of the ship limits one's movements, and the watch pattern of four on and eight off add an expectancy and normality to life on board. Eight bells has just struck, the change of the noon watch. The watches rotate and John, the third sparks, will have relieved the chief by now. I wander down to the wardroom where lunch is being served. Maybe after lunch I will write a letter home.

An air letter form, just a single sheet of thin blue paper with lines for folding and a gummed edge for sealing, lies on my writing desk in the cabin. The blank form stares up at me. I have written *Dear Mother and Family*, and the date, *November 29, 1944*, but that is as far as I have got.

Frustrated, I look round my cabin for inspiration. The curtains covering the heavy steel shutter on the porthole are light blue; they match the wall-to-wall carpet. A fan perched high up in the corner of the bulkhead watches me over the large bunk. A settee, cupboard and washstand complete the picture. This is my new home; the drawers under the bunk and the cupboard are filled with all my worldly possessions. In the cubbyholes on the writing desk, wedged in against the roll and pitch of the ship, are photographs of my sister, my two younger brothers, and my mother.

There is so much to say, so much to tell, but then there is the censor. I cannot talk about the ship, its destination or anything which can be construed as information useful to the enemy. If I do, then the censor's black pencil and scissors will render the letter useless. My pen scratches the paper surface as I tell my family about my new friends, John and Mike, the shows we saw in London and the dance at the army camp where Mike's sister was stationed.

Brother Isidore's little sacred heart, the talisman he so grudgingly gave me after I was torpedoed in the Atlantic, sits almost forgotten in the top upper right-hand drawer of the desk. I forage in my desk drawer and pick up the little emblem of the bleeding heart. My fingertip gently touches its red satiny surface. It feels soft and memory stirs the anger in my soul, anger against Brother Isidore and the religious order he represents.

My mind goes back to that wintry day in January 1943, and I remember how he stood in the entrance hall of the College and gave me the talisman. Almost two years have gone by and I still have it; it has travelled with me through the North African and Italian campaigns. I reflected that there had been close calls when we crossed the Mediterranean from Algeria to Malta and beyond to Alexandria, but the ship on which I served, the *Empire Harmony*, lived up to its name and came through unscathed. I've grown attached to the talisman, despite my bitterness against the religious order.

The blank page is no longer blank, and satisfied I lay down my pen. Seven bells have already struck and it is now close to four in the afternoon. I glance at the travelling clock on my writing desk; this will be my first watch aboard the *Empire Path*.

It is eight o'clock when I come off watch and mount the stairway to the wheelhouse. The blackout curtains part briefly as I enter and I stand for a moment to allow my eyes to adjust to the dark. The helmsman is silhouetted against the lighter dark of the for'ard wheelhouse windows, and, as I pass behind him, I call goodnight. The faint light blue glow of the gyrocompass illuminates his face, and I see his lips move and face distort in a grin, as he says, "G'night, Sparks."

The wintry air chills my face when I step out onto the wing of the bridge and peer over the weatherboarding. There is no moon or starlight to relieve the dark, only the white wake of the ship ahead; its dimmed blackout navigation lights and the bow waves of ships on our port and starboard beam are visible.

The cold freshens my mind, clearing away the tensions of listening to Morse code through static for four hours. Fortunately, the radio traffic has not been heavy tonight. I draw a deep breath and pull up the collar of my bridgecoat as I lean against the parapet. It's become a habit of mine to come up here for awhile when I come off watch. I find that it relaxes me.

Subconsciously, my eyes take in the dim shapes of nearby ships, and inevitably my thoughts turn to my last leave. A wave of shame and regret sweeps over me as I think of Hazel, and I tear my thoughts away, for the hurt is too deep. *Why, why was I so bloody insensitive that I went along with what my mother said?* I ask myself angrily.

I can't bear the thought of hurting Hazel, or any other woman, come to that. Their tears reduce me to a quivering idiot with a guilt feeling that is hard to bear.

The rhythmic roar and splash of the bow wave comes to my ears as the ship pitches and rolls in the swell of the North Sea. *Yes*, I think, *I'm afraid of women. I don't understand them, their needs, or how I should act. If I did, then I wouldn't get into these messes.* As a product of segregated religious schools, I've missed a normal teenage life. All I've known is the Merchant Navy and war since leaving school.

There's a lump in my throat, and I breathe deeply. The cold North Sea air in my lungs soothes me and my mind turns to other thoughts.

I smile as I think of Mike and John. I like their company; they're both about my own age, and we get on well together. Perhaps that is it, the age difference! On my first two ships, when I was seventeen, my peers were older men. I guess they thought I was just a bloody kid, and treated me as such. But now I'm nineteen, and Mike and John are the same age. Our lives and interests are similar and we can relate to each other, share experiences, talk and laugh together.

A voice breaks into my thoughts. "Daydreaming, Sparks?" It's the bridge officer.

I turn and smile as I recognize the voice; it's the third mate. "Yeah," I say, "you startled me creeping up like that. I was just wondering why I always seem to mess up with girls; I don't seem to understand them!"

He laughs. "Why? Did your girl stand you up?"

"No, I stood her up, and now I can't get it off my mind."

I look up as I sense him grin. "Aaahh," he says. "Women are nothing but trouble. You're young yet. Take my advice. Love 'em and leave them. Once they get their hooks into you, your wandering days are over."

We talk on awhile before I take my leave and return to my cabin. I'll be on watch again at four in the morning.

Sleep does not come easily. When it does come, it is a disturbed and restless sleep, leaving me tired and listless when the watchman calls the next watch. "Come on, Sparks," he calls, "A young man like you should be up and about at this time of the morning." I groan as I look at the clock; it's ten to four. "Here," he says, "I've brought you a nice cuppa coffee, mate. That'll liven you up."

The Dutch coast lies off the convoy's starboard flank when I come off duty and hurry to the wardroom for breakfast. Now, huddled from the cold wind in my bridgecoat, I lean against the rail outside my cabin scanning the coast and dun-coloured sea. Only the steady rhythmic thud of the engines and the constant sound of the waves breaking against the bow disturb the peace. The coast looks calm and serene. Seagulls follow in the ship's wake waiting for scraps from the galley.

The calm is deceptive. Mines litter the estuary, our own and those of the Germans, who nightly sow the channels from minelaying aircraft. With minesweepers sweeping the channels ahead to clear the shipping lanes, the convoy passes through unhindered. In the approaches to the Scheldt Estuary, the formation of the convoy is changing. We do not anchor but, shielded by our own minefields from enemy submarines and surface craft, we slow to half speed and heave to. The naval launch carrying the pilot is already on its way.

My thoughts wander as I watch the launch disappear and suddenly reappear amongst the ships. It reminds me of a small busy bee around flowers, only it is delivering pilots, not pollen, to the vessels as they drift in the sea.

I'm brought back to earth by a dig in the ribs. "Hi, Gord." It's John.

"Thought you were still asleep," I say with a laugh, then nod towards the launch off in the distance. "Just watching the pilot boat. He's busy this morning."

"Anything happen on your watch?"

"No," I reply, "only reports of enemy aircraft over the Scheldt, mainly minelaying aircraft. It's been a relatively quiet night, apart from your report of an E-boat attack."

I'd taken over the watch from him at four this morning. Earlier, he'd received a radio report of an attack by three E-boats, which was beaten off by the escort before it could get to the convoy. One E-boat had been destroyed.

John yawns and stretches. "Guess the minesweepers will be busy today then. Must be a daily routine keeping the estuary and river open."

The approaching roar of the naval launch breaks into our conversation, and we see it suddenly appear around the stern of the ship. A naval officer stands on its deck with one hand gripping a handrail on the side of the launch's cabin. With a practised eye and hand, he catches hold of the Jacob's ladder and climbs up the ship's side. His battered peaked hat appears over the top of the steel bulwark followed by his leg as he climbs over and onto the deck.

A grin spreads over his face as he glances at the little group of men watching his efforts. "Hi, there," he calls, "bloody fine day, mates. Old Man on the bridge?"

The bosun nods and with a quick wave the pilot disappears along the deck and up the steps to the bridge.

The ship vibrates as the propeller churns the sea. Slowly at first, we gather speed and head for the gaping mouth of the Scheldt. As we enter, it narrows slightly as if to swallow us. Our voyage into the unknown has begun, but to the experienced pilot it is routine.

Dressed in battle fatigues, the DEMS army gunner attached to our ship sits in the Oerlikon gun nest. The smoke from the burning cigarette in his left hand spirals up momentarily, then is whipped away in an invisible stream by the early morning breeze. He glances up at me and grins as I watch the smoke. He is not much older than I am. A tuft of fair hair escapes from beneath his steel helmet.

"Nice morning, aye, Sparks," he says. "Maybe we'll get a chance to bash the bastards if they come in low." He laughs and the tension eases.

We are on full alert. Many eyes search the skies for signs of developing attacks. Experience in the North Africa operations and Italy has killed any illusions I might have about the glory of war. If an air attack comes, it will be quick. There will be the sudden deafening roar of low-flying aircraft engines, the chatter of machine guns as bullets and cannon shells rake the decks, followed by the whine and blast of bombs.

When under attack on this ship, I will man the machine gun on monkey island above the bridge, slightly for'ard of the funnel. The standard metal shield reinforced with sandbags will offer some protection, though not much, against the concentrated fire of a diving aircraft armed with cannon and machine guns in the leading edges of its wings.

Standing by the vertical iron ladder on the wheelhouse bulkhead leading to the machine gun nest, I think back to the simulated attack practices on the gunnery course in Southend. It was August 1942 and the grass was green and fresh. On what was once a golf course, twenty or thirty merchant seamen, in a motley array of clothing, wait in line. A machine gun on a tripod stands ten yards away from the nearest man. A small white parachute has been shot up into the clear blue sky. As it descends slowly to earth, the small target beneath sways in the breeze.

A command rings out and the first man in the line dashes to the gun, fires a short, one-second burst, then falls to the ground, simulating death. Pushing him out of the way the man next in line takes his place and fires at the drifting target. Now it is my turn and I dash forward, aim and fire. The target disintegrates, its chute crumples and spirals down in its death throes. The gunnery officer is pleased.

Antwerp

THE FLASH of signal lamps winking in the distance draws our attention to the dim distant shapes of the minesweepers silhouetted against the shoreline ahead. Small, they resemble converted trawlers. The third focuses his binoculars on them, studying them for a moment. "They're preparing to stream their paravanes, Sparks," he says. "It looks as if they're going to make another sweep of the channels ahead of us. Can you read the message they're sending?"

"No. The minesweepers are simply acknowledging messages they receive from the shore naval station, but I can't see that lamp. You're probably right, they're heading into the Scheldt."

The third lets the binoculars fall and they dangle by the strap round his neck. His eyes are thoughtful as his hands unconsciously fumble for a cigarette, "Those paravanes are okay for tethered mines, but they're useless against mines on the seabed." He pauses to light the cigarette, cupping the match against the breeze with his hands. "The Germans brought out a new mine earlier this year. From what I hear at the naval conferences, we are not having too much success against it. Have you heard about it?"

A vague recollection teases my mind. With a shrug, I say, "I've heard something about a ratchet mine. Is that it?"

His eyes on the minesweepers, the third nods. "Yes, it's a pressure operated mine with a ratchet type detonator." Unconsciously, he strokes his chin as he continues. "The minesweepers can sweep over it and it won't explode until the number of passes reaches a preset detonation point."

He draws deeply on the cigarette, inhales the smoke, then lets it trickle out slowly as he exhales. "At one of our briefings, we were told they are called oyster mines. The bloody things sit on the bottom in coastal waters where the seabed is less than two hundred fathoms." He pauses, then adds, "Just hope we don't come across them. It's kinda like playing Russian roulette."

As we draw nearer, I watch the mouth of the Scheldt, studying it through the binoculars. The Scheldt is wider than I had expected. Curious, I wander back into the chartroom to check its width.

The course through the Scheldt has already been drawn on the chart. The estuary is five miles across from bank to bank. Using the chartroom calipers, I also check the distance to Antwerp. It is forty miles through the Western Scheldt to the Scheldt River, then another fifteen miles or so further to Antwerp. All going well, the convoy should reach port in eight hours. With the shortening daylight hours, dusk falls about four or four-thirty, so the last few miles we will have to navigate guided by shaded light buoy markers and the pilot's navigational skill. Satisfied, I retrace my steps through the wheelhouse and out onto the bridge.

The land on both sides of the Scheldt is flat, mainly marshland. The powerful binoculars bring the banks closer and I can make out the broken shapes of beached ships and other evidence of war which litter the banks. As we penetrate deeper into the Scheldt, the swept channel swings wide, and the ship hugs close to the northern bank. I call to the third mate, "Look, Steve, the land beyond the bank's flooded! The dikes have been blown up."

"Yes, I know, but look ahead of the ship. Over there." He points with his arm ahead and to one side. "It's the town of Flushing."

We watch in silence as the ship passes the small deserted town on our port bow. The outskirts and suburbs are under water. In places only the roofs are visible. An old church tower, gaunt and severe, pokes up out of the sea like a flagpole. The town centre and dockyard areas have been heavily damaged by the recent fighting, but the flooding is less severe; some of the land is above water.

The noise of war rumbles to the east and north of us. Ahead, unexplained explosions rend the air, a constant reminder that the war is still being fought. Enemy aircraft are markedly absent, but there are mines still underfoot, waiting for the unwary.

Mines! I do not envy the engine room crew and stokers. They don't stand a chance if a mine explodes beneath the ship. Only a thin sheet of metal, maybe half an inch or so thick, lies between them and the waters of the Scheldt. However, it's not that much better for them in an air attack. Entombed below, they can only sweat it out and pray that the ship will survive.

I think it's preferable up here on the bridge. At least I can see some of the action, which makes me feel better. There's also the machine gun; I like the feel of it in my hands. It may be useless against fast-travelling aircraft, but I feel I can hit back and am not just a sitting duck waiting tamely to die.

Unchanging flat flooded land, interspersed with war-damaged and half-submerged buildings, stretches away to the north. The Scheldt twists and turns, first north and then southeast again as we keep station behind the ship ahead of us.

Rounding a bend, we suddenly come upon a wrecked cargo ship listing at a drunken angle, beached, half on and half off a sandbank. Like a dying animal, the ship has managed to drag itself away from the narrow shipping channel. Smoke still drifts from its broken body, and a flicker of flame flares in the stern crew quarters. The ship appears deserted. We can see no movement on board as we scan her with our binoculars.

I turn to Steve, who is silently watching the vessel as we pass on down the channel. "She must have been hit just a few hours ago. What do you make of it? Did she hit a mine?"

Steve continues to look at the vessel for a moment without answering, then says, "Yes, I think so. The engine room cowling is peeled back and her back is broken. With damage like that it's more than likely a mine, probably a ratchet type which the minesweepers didn't find."

He pauses, stretching against the bulwarks, then relaxes and continues. "At least they managed to beach her clear of the channel, so she's not blocking shipping. It must have happened today because she's still well alight. She may have been from our convoy. The lead ships passed this point some time ago."

It is late in the afternoon, around four o'clock, before we reach the Scheldt River. The light is fading in the overcast sky as the pilot brings the ship round hard to starboard. Close on its tail, we follow the disappearing stern of the cargo ship ahead as it leaves the broad Western Scheldt and enters the river on the last leg of the journey. The Dutch frontier post goes unnoticed as the ships cross the border. We are now in Belgium, and Antwerp is a scant fifteen miles away.

Broad and slow-flowing, the Scheldt (or Escaut) River is navigable for most of its 270-mile length. It originates in northern France before winding its way through locks and towns in Belgium to the Scheldt Estuary and the sea. Our convoy passes slowly along the historic waterway. In the gathering dusk the shaded green and red bridge lights of the ships follow each other in single file spaced evenly a few ship lengths apart.

With the coming of night, the captain and pilot talk in low, tense voices as they check their bearings on the dimly-lit buoys marking the channel. The shadowy figure of the helmsman stands motionless by the wheel, as if in a trance, his eyes fixed on the lighted compass in front of him. The illuminated dial shows the ship's heading and casts a gentle soft glow in the darkened wheelhouse.

The pilot's calm, confident voice calls out a course correction, "Starboard five points, steady as she goes." The command is followed by a swing of the wheel as the helmsman alters course, his eyes on the compass needle. The ship's bow responds to the turn of the rudder and the anchored buoy slips by on our port side.

Two hours have passed by, and now little tugboats fuss around us. Their presence is apparent in the light glades on the river surface from the dimmed blackout lights of the dock area. Their silhouettes break the glades as they come with gleaming red and green eyes which bob up and down, swaying in the ship's wake.

The bridge telegraph handles creak and clatter as the pilot pulls them back a few notches. The telegraph registers *half speed ahead* and is followed by the answering ring from the engine room as the speed drops off. Another urgent ring signals *full astern*, and slowly the *way*, or forward drift, of the vessel is halted as we heave to. The bustle and shouts of the bosun and deck crew break the quiet of the night as ropes are cast and the tow lines made fast. The *Empire Path* is in the hands of the tugs and the skill of the pilot. The ship edges gently into our berth alongside the dock. We have arrived in Antwerp. The log on the bridge shows our docking time as 1830 hours, six-thirty in the evening.

Air raid sirens, followed by the familiar *chug, chug* of doodlebugs, brings me out on deck that first night in Antwerp.

"Sound familiar, John?" I ask him, as we watch the jet flames from the ramjet engines streak across the night sky. We're together under the shelter of the bridge.

John grins and says, "You can't get away from the bloody things, can you."

Instinctively, we shrink back as an engine cuts out over the city. The flying bomb glides down silently, and seconds later there is a blinding flash in the night sky. The explosion rolls over the city and docks. One after another, the ramjet engines cut and the flying bombs glide down on a path of death and destruction. Bursting anti-aircraft shells splatter the sky around the doodlebugs, which travel in unswerving lines at close to four hundred miles per hour.

In a pyrotechnic outburst that turns the night into a fascinating, but deadly, fireworks display, the tracer shells from Bofor guns in the docks and surrounding sites open up. They fill the night sky with streams of tracer shells that, from a distance, seem to reach up lazily, almost floating up into the sky in concentrated arcs of fire.

An approaching doodlebug cuts its engine. This one's coming close and we hit the deck, hiding behind the thick metal bulwarks. There is an indistinct whining sound and an ear-shattering explosion when it bursts in a nearby warehouse.

"Christ," says John as we get to our feet. "That was too bloody close." The distant sound of fire engines and emergency vehicles racing to the scene reaches our ears. Rescue crews are already at work, sorting through the wreckage.

At first light the following morning, military personnel and local dock workers swarm aboard the *Empire Path*. Tarpaulins and hatch covers are stripped off, the hum of winches and the squeaking and creaking of straining blocks and tackle add to the din as the long process of unloading cargo begins.

Three weeks, three eventful weeks we are in Antwerp. Day and night, enemy forces bombard the city and dock areas with doodlebugs, V2 rockets, and, occasionally, enemy night bombers.

The undulating wail of air raid sirens warns us of incoming enemy aircraft or doodlebugs, but there is no warning against the V2. Supplies are vital, and between raids the unloading goes on. The deck officers are busy supervising the crew and dock workers unloading the cargo. For the radio staff, apart from maintenance duties, our time is our own.

Sometimes I wander into town alone, but mostly I'm with either Mike or John. Transport's no problem. There are always army jeeps, Landrovers and trucks going to and from the docks. Hitch hiking is the order of the day. Today's no different. It's mid-December; we've been here a little over two weeks now. Mike's with me, and John's gone for the ship's mail in an army post office in the docks some distance from the ship.

Snow is falling on the street outside, but it's warm in the cafe where I sit with Mike at a small windowside table. It's too damned cold to wander long in the streets of Antwerp; besides there's a nice young waitress in here. Mike talks to her when she comes to take our order. His French, like mine, is limited to the few words learned at school. As he speaks, he brushes back a lock of fair hair that's fallen in his eyes.

Her eyes twinkle. She's used to servicemen, and plays along with him as he kids her. "*Oui, monsieur, café. Et vous?*" She looks at me.

I smile, then carefully choosing my words, I say hesitatingly. "*Moi aussi, la même chose,*" then add in English, just to be sure, "The same, please."

"Nice bit of stuff that, aye, Gord," Mike says as she turns and walks away towards the back of the cafe. Our eyes follow her swaying hips and slim legs.

"Yeah, you're right. Pity we'll be going soon."

The coffee's strong and I sweeten it with sugar; Mike likes his unsweetened and black. *Yuk*, I think, *strange taste*. We chat awhile about the girl, she's got big boobs and a low neckline. At least we both have similar tastes in women. But it's a lost cause; there're too many servicemen around, and she can have her pick. We're here today and gone tomorrow. A lot of these men are based here, back area wallahs mostly.

Our talk turns to more urgent matters. The Old Man's put the ship on alert by order of the governing military authority.

"Think they'll break through, Mike?"

"I don't know. It's hard to say; we are getting so many rumours now. The Americans are holding the Ardennes, and, if they break, then we've had it. The German tanks will get through to the coast and cut us off. That's what the Old

Man said, didn't he?" Mike shrugs his shoulders in resignation. "We will just have to sit tight and see what happens."

"Yes, I guess so." I half nod and add, "If that happens, we'll probably end up in a prison camp for the rest of the war. Not a pleasant prospect, is it?"

A light covering of snow powders the sidewalks, but the paved streets are clear when we leave the cafe and walk past the large department stores of the downtown area. This city had once been a thriving commercial centre, but it presents the familiar face of all occupied towns under military control. Soldiers—Canadian, American and British—are everywhere and mingle with civilians hurrying to and from work or simply shopping. Bomb-damaged buildings, some still smoking from recent hits by V1s or V2s, are a constant reminder, if one needs it, that the front line is very close.

We are in one of the main streets when Mike stops suddenly. "Gord," he says, "look, this must be where the Rex Cinema was." He points. "Over there, there's still some of the movie placards standing. You must have heard about it. It was hit by a V2 rocket three or four days ago."

"Yes, the mate was talking about it in the wardroom. He said hundreds of servicemen and civilians were killed and almost as many seriously wounded. The cinema was crowded at the time." I gaze at the rubble and ruins of what had been a major movie theatre. It is gutted and still cordoned off. I wonder if they have got all the bodies out. "Poor devils," I murmur as we move on. An army truck passes by going towards the docks, so we signal the driver and hitch a ride.

Unloaded and with ballast and some cargo aboard, we prepare to return to England. At first light on December 24, the hawsers holding us tight to the dockside are loosened and cast off. We are free to move slowly out into the river. With the pilot aboard, the *Empire Path* leaves Antwerp behind and makes her way downriver into the Western Scheldt.

Outward bound to the sea, we pass more wrecked ships. Some have been hit by mines laid by German aircraft; others were the victims of enemy bombing attacks. Some are still burning; smoke billows up into the leaden overcast sky as they lay helplessly beached on the shore. It is mid-afternoon before we reach the estuary of the Scheldt and the tension seems to lift. The pilot boat turns in a racing sweep away from the ship's side, carrying the pilot back to his base to await the next incoming convoy as we head out to the open sea.

Death of the *Empire Path*

THE STATE OF ALERT ends when we emerge from the estuary; the narrow confines of the Scheldt and constant threat of air attacks are behind us. Nothing lies ahead now but the North Sea. The mood aboard ship changes. An undercurrent of excitement adds to the festive air; we are going home. It shows in the faces of the officers and crew; they smile more, are less terse in their manner and conversation is spontaneous.

The chief is on watch in the radio cabin. I'll relieve him in an hour's time. John is beside me on the quarter deck as we watch the pilot's boat disappear in the distance.

The ship's engine takes on a new beat as the *Empire Path* surges forward. Flags are flying from the mainmasts, and there is an air of purpose, almost eagerness, as the ships of the convoy jockey for position.

John turns away from the ship's rail, leans back on it and looks up at the bridge. His eyes are happy and his features relaxed. "Just think of it, Gord," he says. "We'll be in the Thames Estuary by tomorrow night, and, if we're lucky, in dock by Boxing Day." He chuckles and says, "Man, we'll make it for New Year, that's for sure. Mum and Dad will be pleased."

Laughing, I agree with him as we part. It's cold out here and there's an icy wind blowing. The distant shoreline is white. It's winter in Europe and the Dutch and Belgian coasts are encrusted with snow. I seek the warmth and comfort of the ship's interior.

It is Christmas Eve and the wardroom is decorated with colourful paper chains, bells and other ornaments. We even have a Christmas tree; the chief steward must have got it from the ship's chandler in Antwerp. It stands in the corner ablaze with miniature lights, reminding me of other Christmases.

Restless, I wander past the galley. The tantalizing aroma of food assails my nostrils and awakens my appetite. There will be a special dinner tonight. Curious, I glance in the doorway for a moment to watch the cooks busy preparing the evening meal.

One of them sees me as I poke my head around the galley door. "Hi, Sparks," he cries, "Feeling hungry already?"

"Smells real good." I say with a laugh. "What is it?"

"That's a cook's secret, a bit of this and a bit of that. You'll just have to wait for dinner. It'll taste better, mate." He turns and laughs as he says something to the others, all busy with various chores among the steaming pots and pans. I wave and pass on.

Mounting the stairway from the galley, I walk along the warren of passageways in the midships accommodation, and emerge onto the deck. The air is chilly and cuts through my uniform jacket and sweater as I stand for a brief moment by the starboard railings. We are free of the Scheldt, and with other ships ahead, astern and abeam of us, I feel secure. Tomorrow we will be in the Thames Estuary. With hardly a backward glance I hurry to my cabin on the upper bridge deck.

It's almost four and the ship's bell is ringing the change of watch. I wash my hands and face before going to relieve the chief in the radio cabin. Straightening my tie, I take a last look in the mirror and turn to leave the cabin, but I never make it to the cabin door: a deafening roar catches me in mid-stride. The blast tosses me into the air, slamming me against the bulkhead of the cabin with stunning force. The sensation of falling is tinged with fear, fear of falling into a bottomless pit in the bowels of the ship. My stomach churns and time stands still.

Stunned, I land on my hands and knees in the debris of what had been my cabin. *I can't see. It's pitch black. Am I blind?* I shake my head to clear the ringing sensation in my ears. Then, as my senses return, I can see. A half sob of relief catches in my throat. *I'm all right.* There is a faint glimmer from the opening where the door had been; the emergency light in the passage is on. The door's blown off completely or is hanging on its hinges; I don't know. The ship's generators have been knocked out in the explosion; all electrical power, except the emergency batteries, has gone.

The acrid smell of explosives, smoke and dust add to the confusion, but it's the vibrations and cracking sounds exploding like pistol shots that scare me. The ship is shaking violently as her forward speed is checked by the sea pouring into her ruptured hull. She is breaking up and could go under any instant.

The strength of the sea water tears and rips the metal plates in a cacophony of tortured shrieks as they grind against each other and rip off. The noise is amplified in the closed interior and sets up a mind-boggling, unearthly din against the rushing sound of water in the ship's bowels.

Strange! The fear is there. I can feel it in the tightening in my throat, but I do not panic; I make my way out of the cabin and up the dim passage to the beckoning patch of daylight, which leads to the wheelhouse. Thoughts race through my mind. *Mined or torpedoed?* I don't know, and care little, only knowing that I must get out to the deck or be trapped in the sinking ship.

Pausing at the door of the radio cabin, I glance in. The emergency light is on, but the cabin is empty. There's just twisted wrecked equipment. *The chief's gone!* With renewed haste I scramble through the passage to the wheelhouse. It, too, is empty. *Where is everyone? Have I been unconscious for a few minutes?* Anxiety gnaws in my guts as the fear of being left behind spurs my feet through the wheelhouse door and onto the bridge.

A feeling of relief floods through my body as I find the others. On the boat deck below me, a deck officer and lifeboat crew are preparing to lower the port midships lifeboat. Silently I watch as they swing the lifeboat out over the side on its davits. Momentarily, my eyes rove the ship's superstructure; the engine room cowlings are curved open, shattered by the force of the explosion. It looks as if a giant fist had punched right through the ship, because the funnel, too, is partly toppled. The ship's boilers must have exploded at the same time that we were hit.

The creak and groan of the dying ship, already splitting across the mid section, lends speed to the boat crew's efforts. I look away for alternative escape routes in case the ship suddenly plunges into the depths. The ship's bow is rising up in the air, and I have the feeling that the bridge is folding into the fore deck. The vessel gives another sickening lurch and settles in her middle. It is a question of just how long she will hold together before breaking in two.

From the height of the bridge, I see a small group of seamen on the ship's bow, maybe ten at the most. They are clinging to the rails, watching a minesweeper cautiously approach. As her bow noses alongside the starboard bow of our ship, the seamen scramble down ropes or jump onto her deck. They are safe, and the minesweeper, reversing its propellers, pulls clear.

As I turn away from the bridge bulwark, I glance back. The lifeboat is being lowered to deck level, and the deck officer yells at me, "Come on, Sparks, get in. She may go at any moment. The old bitch is breaking her back."

Not hesitating, I step aboard as it is being lowered. Two crewman, one at each end of the boat, stand by the davits, checking the rope running through the pulleys. We hit the water with a splash. The boat rocks violently with the impact, but the seas are moderate and it settles down quickly.

Shouted orders from the deck officer on the boat deck above brings instant response, and painters, as the ropes are called, are made fast fore and aft. With the lifeboat secured against the ship's side, I help the crew hold the bobbing boat away from the ship's side with a boathook until the stuffed canvas buffers are put in place. The roar of water pouring into the ship is louder now. It's a constant background roar interspersed with the shriek of stretching, breaking metal as the ship lurches and settles in the sea.

The scrambling nets shake and jerk spasmodically as men climb down, some silent, others cursing as they step or jump aboard. Memories flood back as I sit there on the lifeboat's deck planking, memories of two years ago, almost to the day. I have seen it before, only now it is daylight, not the confusion of night. We have not been left to fend for ourselves in a hostile North Atlantic, a thousand miles from shore in winter gales. Here, in the North Sea, help is at hand. The captain boards the boat as the boat crew operating the davits slide down ropes and nets to join us.

The captain raises his voice above the noise of the dying ship. "This is the only lifeboat left, men," he says. "All the others have been destroyed in the explosion, including those on the stern. We'll wait one minute to make sure there's nobody else left aboard, then we will have to go. She won't last much longer."

It seems a long minute waiting there, but there is no sign of life. All who can, have probably abandoned ship. The captain nods to the coxswain, "Cast off." The boat crew untie the painters holding us to the ship and slowly the gap widens as we drift clear of the stricken vessel. In the confusion I've lost contact with John and Jim, the chief sparks. There are so many in the boat that I can't see them.

I watch the *Empire Path* settling lower in the water, the bow and stern rising in the air as her centre section sinks. The strong arms of the crew strain at the oars, pulling us away from the ship's side. I sit there in the crowded boat obsessed with my own thoughts. It is like a bad dream, a dream world. Then I hear the captain's voice as he talks to the coxswain in the stern. "Stand off well clear of the ship, about a hundred to a hundred and fifty yards," he says. "We don't want to be pulled down with her when she goes."

The thought of the vortex, as a ship plunges to the depths, spurs on the oarsmen in their efforts. They know its tremendous force as, like a whirlpool, it draws down the loose wreckage from the sinking vessel, only to release it with an equal force back to the surface, killing any living thing in its way. The oarsmen sit two abreast, four on each side, and bend to their task. The oarlocks creak and groan as the oars dip deep into the sea, and move the crowded boat through the rolling waves. Gradually we put distance between us and the sinking ship, until the captain feels that we are clear of danger. “We'll wait here,” he says, his voice despondent. The oarsmen rest on their oars, and our eyes watch the stricken ship while we wait to be picked up by an escort vessel.

“Captain!” a voice calls out excitedly. “There's someone on the stern. Look!” He points towards the ship.

Two figures emerge from the stern crew quarters and dash to the rail. The stern is tilting at an acute angle and the propeller, locked in position, is almost clear of the water as they jump into the sea. The weather is bitterly cold. I shiver at the thought of the freezing water as they flounder in the sea. The memory of the other sinking lifeboat, two years ago in the Atlantic, and the shock of icy water on my body is vivid in my mind. *Poor bastards*, I say to myself.

As we start to row towards them, their frantic shouts reach our ears, but an escort vessel has already seen them and is bearing down on the two heads bobbing in the icy water. We relax as they are hauled aboard the corvette. They are safe, rescued before hypothermia has a chance to set in.

The cold bites through my uniform jacket, but I don't seem to feel it. Events are happening too fast and survival is uppermost in my thoughts. The sound of engines distracts my attention as I sit there absorbed in the unfolding drama. I look hurriedly around and see the curved bow and roundels of an RAF Air-Sea rescue launch approaching at high speed, riding the crests of the waves. Spray shoots in sheets over the bow as the launch slows and falls into the wave troughs. The twin-screw converted motor torpedo boat jams the engines into reverse and sidles up to our lifeboat.

A voice calls out, “Bloody fine time to go boating, mates; 'ere, catch this fucking rope.” Standing on the heaving deck of the launch, the crewman throws the rope across the lifeboat's bow. Lashed fore and aft to the rescue vessel, we waste no time in abandoning the lifeboat and scramble onto the launch's slippery wet deck.

The launch captain's voice comes over the loud hailer, “Have any more lifeboats been launched from your ship?”

Our captain answers, “No, we're the only lifeboat left. All the others were destroyed in the explosion. Some of our men were taken away in a minesweeper, and others were picked up from the sea by the escort.” He makes his way to the bridge of the launch and talks to the other captain.

Men are everywhere, crowding the small for'ard and aft decks, wherever a foothold can be found. In deference to our feelings and our captain's request, the launch waits, hove to in the sea, as we watch our ship's final death throes. Split in two halves, the heavier stern section dives first; then with a defiant last thrust of its nose, the bow disappears in a frothing foam as it, too, slides down beneath the grey murky waters of the North Sea.

I see her go with mixed feelings. There is a sense of loss, and insecurity. My cabin and home, the things one collects, including the talisman, are all irretrievably lost beneath the sea. I look around me. Apart from the ten men picked up off the bow of the ship, and the two rescued from the sea by an escorting corvette, these are all that are left, between thirty and forty survivors.

Where is Mike? I haven't seen him. John and the chief are here. The chief's in a bad way. Looks like a broken shoulder. John's okay. I gaze moodily at the sea. How many men have we lost? The walking wounded, blood seeping through hurriedly bound strips of cloth, are helped by their mates to protected inner areas or laid on the deck.

There's a naval doctor aboard and he's attending to the more seriously wounded men. A man, one of the galley crew, sits near me, moaning softly; he's badly scalded on face and hands. It's all in a haze; my mind's not really taking it in. There is just an awareness of jumbled bodies nursing broken limbs, their faces chalky white and haggard with shock.

I look around for Mike but still can't see him. I hope to Christ he was not on watch in the engine room when the explosion tore right through it; the engine room crew on duty could not have survived. I shake my head; the poor devils didn't stand a chance.

Its deck crowded, the launch prepares to leave. The ship has gone: only floating debris remains on the surface of the sea. We are a motley crew in various stages of dress, most in dirty, torn, everyday clothing. Some wear life jackets, but others, like myself, are without one. I could not find mine amongst the rubble of my cabin. Now, sitting on the deck sunk in my own thoughts, it is the least of my worries. On the seabed of the Scheldt Estuary, lie the ship, its dead, and the talisman.

The Aftermath

THE LAUNCH turns in a slow arc to the south; then, at full throttle, it surges forward at thirty to forty knots, leaping from wave top to wave top en route to its base in Ostende. As it thrusts forward, the cold biting wind and salt spray thrown up by its passage drive us to find whatever shelter we can from the freezing weather.

As I sit huddled up with others on the inside deck of the casualty clearing section, an RAF crew member hands me a lighted cigarette. "Here, mate," he says, "it'll make you feel better." I puff experimentally, coughing when the unfamiliar smoke enters my lungs. Unperturbed, he continues, "Settles your nerves." Then, as if in an afterthought, he adds, "You know, this must be a record number of survivors for us; there's a good thirty men aboard."

I look up at him and smile; the smoke tickles my throat. I have not tried a cigarette before, and the taste and smoke bother me. However, I persevere and smoke it so as not to offend him.

He chats amiably to us as we sit in a group around him. His voice rises to combat the noise of the engines and the incessant drumming as the launch's hull skims the wave tops. "Yes, mostly we pick up crews of ditching bombers." He pauses and gestures with his right hand. "There are several air-sea rescue launches stationed out here off the coast. Usually we pick up five or six aircrew, sometimes more from the bigger bombers, but now! Christ, just look at this lot." One of the launch's crew beckons him from the open hatchway, and, scrambling to his feet, he makes for the deck.

The motion is getting to me and I feel the onslaught of seasickness. I have to lie flat, or I will be sick. I manage to find a space close to the bulkhead. No longer interested in the world I lie there, face down, clinging to the bucking deck.

A voice arouses me from my stupor, penetrating the fog of sickness that surrounds me. It is a doctor kneeling by my side. "Are you okay, son?" he says.

"Yes," I reply and turn my head to look up at him. "It's only seasickness." Then add, "It helps if I lie down."

He laughs and says, "Yes, it will. We'll soon be in Ostende, then you'll be all right." He rises quickly and turns away to attend to others needing his help.

The rescue launch slows and enters the harbour mouth in Ostende some twenty-five to thirty minutes later. Relief sweeps over me as the incessant drumming and bouncing of the launch gives way to the smooth glide of still water under its hull. The world stops shaking. I get to my feet and stand by the boat's railings.

As the launch makes its way to its berth, Ostende harbour lies before us, grey and unkempt, but with a light mantle of snow covering the ground. Disembarking in the growing twilight, we assemble on the dockside. RAF trucks with red crosses are there to take the wounded to hospital. I see Jim, the chief sparks, as he steps into the ambulance, but he is hurried aboard before I can speak to him. His arm is in a sling and his usually florid features look shaken and white. Blood oozes from several facial cuts.

I ask a nearby medical orderly how badly he's hurt. "The officer?" he says. "The big fellah that's just gone in the ambulance?"

I nod. "Yes, that's him, Jim Hankley, my chief."

"Oh, don't worry, mate," the orderly replies. "He's been hit by flying debris, but apart from a broken collarbone, superficial cuts, and shock, he'll be okay."

I thank him and return to my group standing farther away on the dockside. The uninjured are split into two groups of officers and men. John stands beside me with the ship's officers. Suffering from the aftereffects of shock, I shiver in the raw cold that eats through my clothing. The mate and an army officer stand a little away from us, out of earshot. Stamping my feet to keep warm, I glance towards them every now and then. The mate is examining a list in his hands and is talking earnestly with the officer.

Finally he steps back and turns towards the two groups waiting on the dock. "The bosun, Mr. MacPherson, will be in charge of the crew for billeting purposes," he says, then adds, "The sergeant here will show your group the way to your quarters." He half turns and looks directly at our group. "The rest of you will follow the lieutenant and me to our billet. We will be staying in the army transit camp in the docks until passage can be arranged to get us home."

Thankful to be on the move, our straggly line moves into the honeycomb of warehouses. Railroad sidings and bomb-damaged buildings litter the interconnecting roadways with rubble. The outline of a large barnlike building looms before us as darkness falls over the docks.

We have arrived, and follow the mate inside the building. It smells strange to my nostrils. The mustiness of mildew combined with the smell of straw permeates the building, but at least it is warmer here than outside. Hurricane lamps, hanging on beams, cast a fitful light. The rays are too weak to reach the inner recesses of the building, but the lamps, at least, are a familiar sight, and remind me of home.

The tall shadowy figure of the mate moves to a cleared space in the centre of the building. We trail after him in small groups and gather round him when he stops. As the captain is not here, the mate is the senior officer present. There is a list in his hand and we are anxious to know the fate of our friends on board ship.

The yellow light of a hurricane lamp flickers on the faces of the ship's officers grouped around the mate as he reads aloud from the casualty list. I listen intently to his words as he says, "The engine crew on duty in the engine room were all killed." He pauses, then continues, his voice seems unnaturally loud as he reads the names, "Mike Farland, junior engineer, dead. Ernest Halliday, greaser, and Bert Osmond, stoker, dead." The mate hesitates momentarily and then lists the wounded taken to hospital, the chief amongst them.

I am not listening now; my throat dries up momentarily because my worst fears have been confirmed. I sit down on a nearby box, not trusting myself to speak. *Mike's dead.* The words hammer into my brain. *It can't be.* Jim Hankley, the chief sparks, is in a hospital bed somewhere in Ostende. The mate puts down the list, and an undertone of voices breaks out as the men disperse and talk amongst themselves. *Nineteen,* I think bitterly. *Mike was too young to die.* I turn to John, the third sparks. "He's gone, John," I mutter, more to myself than to him.

"Yes, I know, Gord; it's official now. That's the latest list. He couldn't have been picked up by another ship." John's voice trails off, because he, too, is stricken by Mike's death. He looks weary, but then so do we all. His grey eyes seem to have lost their sparkle; even his longish fair wavy hair, usually swept back, is unkempt.

Half listening to the murmured conversations around me, I am too tired and emotionally drained to comment. The experience of the last few hours has numbed my senses. Shrugging my shoulders to relieve the tensions in my neck, I look around at the large straw-filled wooden boxes that will be our beds. In little groups we sit or sprawl in the straw. Our uniforms are torn and dirty from the smoke and soot of the explosion that literally blew the ship in half.

As the mate sits leaning forward on the edge of a box, the hurricane lamp throws a shadow of his face on the bare wooden wall. A tall man in his late thirties or early forties, he turns his head as I approach him.

"Mr. Mate," I say hesitatingly, "Mike, he was my friend. John, Mike, and I used to go out together. We met his sister; she's a WAAC in an army camp near the Tilbury docks."

"Yeah, I know, Sparks," he says in a low voice. "I'm sorry, but he was on watch in the engine room. None of them survived; they didn't stand a chance."

Standing there a moment, fidgeting with the congealed blood from a small cut on my forearm, I ask, "Was it a mine, sir? I heard it was from other officers here."

"Yes, it was a mine, a ratchet mine. It blew up directly under us, and blew the boilers at the same time."

He notices that my fingers are nervously playing with dried blood on my arm. "You got a cut there?" he asks.

I nod, "Yes, but it's only a scratch. It bleeds when I knock it."

"Well, I'd better put a piece of gauze and tape on it." He looks up and calls out. "If anyone else has cuts or bruises, come over here. I'll get the first aid box."

Soon after the mate has closed his hurriedly organized clinic, an army sergeant summons us for the evening meal. In a straggling line we follow him into the night. A light covering of snow blankets the ground, and it crunches in the hard frost as our feet tread its surface. The night air is freezing and a chill wind nips my nose and hands. There are no shaded lights, for the blackout is strictly enforced in the docks. I keep close to the man in front of me as we follow the dim outline of the sergeant.

In the cookhouse we are each issued with a tin plate and mug, but no knives, forks or spoons, and join the line of men filing past the cooks. Puzzled by the lack of utensils, I ask the man ahead of me where they are. He is an army man and grins at my question. "This is an army transit camp, mate. We carry our own billy cans. They are a part of our kit and come with knives, forks, and spoons. You're lucky they scrounged up them plates and mugs."

The line slowly shuffles forward, and now it is my turn to be served.

"Newcomers!" grinned the cook. "Welcome to 'Otel Ostende. 'Ave some grub, mate." So saying, he wallops a big dollop of mashed potatoes, meat, and vegetables on my plate.

Back in our transit quarters, we sit in the straw and eat the evening meal with our fingers. I'm hungry, and, cleaning the remnants of food off the plate, I lick my fingers. *It is a long way to go to the latrines, and where is the washroom?* There don't seem to be any facilities except a standcock in the yard outside the building.

My eyes close with weariness. Stress and the events of the day are catching up with us all, and, as if by mutual consent, one by one we turn in. Pillowing my head in my arms, I try to make myself comfortable on the straw bed. Giving in to exhaustion, my body relaxes and I fall into a deep sleep.

Christmas Day dawns on a cold and white Ostende. I spend the day wandering around the battered docks with its railway tracks, engine sidings, and warehouses.

Strangely enough it is not the look of the buildings that remains most vividly in my mind, but the feel and crackle of the frosty ground underfoot. A little snow has fallen and the cold air freezes in my lungs as I breathe in deeply. That evening there is no special dinner, or even decorations, to mark Christmas in the canteen when we line up for the limited fare provided for the camp occupants.

Early the next day, December 26, our embarkation orders are received for the short voyage across the North Sea. We are to be transported by a Royal Navy LST which is moored alongside the docks. It is a long flat-bottomed vessel crowded with troops, mainly Americans who are either returning to Britain for leave, or onward-bound to the States.

Our small group is headed by the mate. I look around with curiosity as we board her, then glance at John walking by my side. "She's a strange looking ship, John, isn't she?"

"Yeah, where'd you think they'll put everyone?"

"Guess we'll know soon." I nod towards the bridge. "The bridge's aft, and that's where the main accommodation is. I imagine the troops either sleep in their vehicles or in bunks along the sides of the ship."

Our conversation is interrupted as we file through a bulkhead door. Then we enter into a network of passageways in the stern of the vessel. A Royal Navy officer leads the way to the wardroom, where sleeping accommodations are being worked out.

Whether there is room in the officers' quarters or wardroom, is debatable. Three of us—John, a young engineer, and I, all junior officers—are told there is no room. We are left to wait outside in the passageway while alternative arrangements are made. Eventually, we are assigned to tiered bunks in a narrow passageway. Although crammed with similar bunks and army personnel, this is a definite improvement on the straw-filled hay boxes we had in the transit camp. I have, however, a feeling that it's the age-old class distinction and snobbery of the Royal Navy officer class in operation.

This feeling is reinforced when the junior engineer is quickly transferred back to the wardroom, leaving John and me on our own. His senior officer has spoken up for him: the problem is that Jim, our chief, is in hospital, and we have no one to speak for us. We feel deserted by our former colleagues just when we need the companionship, support and understanding of these men; men who are now, no doubt, carousing in the officers' mess, talking of their experiences, while John and I are left alone to wonder why! Cut off from our former peers, we do not even dine with them, but line up in a mess hall for our food.

When we first line up, a petty officer speaks to us. "You shouldn't be here," he says. "You're officers, there's been a mistake. Don't worry, I'll sort it out. We'll go up to the wardroom." He does his best, but the answer is equally brief. *No, we're sorry but the mess room is full.*

Controlling my anger, I shrug and turn away. *If they had wanted to squeeze us in, they could have done so. I don't believe the navy officers, but what can we do? It is a Royal Navy ship, and they have the right to refuse.*

Embittered, I think of Jim, our chief. *This wouldn't have happened if he'd been here.* Illogically, in my subconscious I blame him, much like I'd blamed my father for dying when I was young. Neither wanted to be wounded or die, but my thoughts are irrational. I feel humiliated and shamed by the insensitivity of my peers. *After all we've gone through together, they don't even try to contact us and ask how we are.* Perhaps they are ashamed of their actions. I hope so, because I will always remember them with bitterness.

Normally, this would be an overnight journey. However, we lie fogbound in the Thames Estuary for two days, long frustrating days as we peer from the ship's rails waiting for the thick fog to lift. At last on the morning of the third day, the fog clears and the order to proceed is given to the convoy. Slowly, in single file, the ships make their way upriver to the disembarkation docks in Tilbury. It is now Wednesday, December 30, 1944.

The debriefing process is short and simple. I've lost my bridgecoat, uniform cap, and all my personal effects. All I have is the soiled uniform I'm wearing. We are directed into another room and given civilian clothes. The weather is cold, damp, and raw. The chill bites through my clothing and I'm glad of the coat.

The representative of the Radio Officers Union asks if I have any complaints. I look at him. *Complaints?* The bitterness wells in my throat. I can't find the words to articulate my feelings. They suddenly seem so petty. A feeling of depression surges over me. *What the hell does it matter any more.* I shake my head and mutter. "It doesn't matter. I just want to go home." I leave him there talking to John. In retrospect, maybe I should have made an issue of it. It would have released some of the anger building deep in my mind.

Unheralded, I arrive home late that night, walking in on my mother and family without any baggage to call my own. I have a shopping bag with a pair of pants, underwear and a shirt or two, part of the survivors' kit issued by the Ministry of War Transport.

On Leave, 1945

"MR. MUMFORD!" The company personnel manager stares at me incredulously. "But . . . but this, I can't believe! We've just received a telegram from the government saying that your ship's been sunk."

I can't suppress a grin at his consternation as he fusses around and offers me a chair. "Yes, that's true enough," I say as I sit down and wait for him to settle himself behind his desk. "We docked in Tilbury last night on an LST from Ostende. I went home, and then caught the first train up to town this morning. I thought I'd better report, as the chief can't; he's injured."

He sends for coffee, then says, with a note of concern in his voice, "Jim Hankley's injured? Oh no! Where is he?"

"Ostende. He's in hospital with a broken collarbone. Shaken up, but apart from that he appears okay. We were picked up by an RAF air-sea rescue launch." I pause as my mind goes back to the scene on the dockside, then add quietly, controlling my emotions, "The last I saw of him was on the landing dock in Ostende. He was sent off to hospital in an ambulance."

His face is serious as he says, "I'm glad you've come because I was on the point of sending out telegrams to your families." He looks down at the papers before him, deep in thought, then continues, "Now I can tell Jim's family the news; they'll be relieved. What about John? Is he all right?"

I nod. "Yes, John should be home now. He left Tilbury the same time as I did last night."

The manager gets up and excuses himself. "Hold on and drink your coffee. I'll be back in a moment; I must get that telegram off right away."

An elderly man in his late forties, the personnel manager is the fatherly sort and comfortable to be with, the type of man you open up to. I like him; he has an understanding manner about him. I wonder how many other young men have sat here over the last few years, telling him their stories. It can't be an easy job, breaking bad news to employees' families, especially where death is concerned.

There's a smile on his face when he returns. "Well, that's that. Now you say an RAF air-sea rescue launch picked you up? Where were you sunk?"

Sipping the hot coffee I launch into my story. He listens attentively, interspersing my report with comments and questions. He's an old seaman himself and has an in-depth knowledge of the ways of the sea.

It's noon before my report is completed. "How about some lunch," he says. "There's a nice little pub around the corner from here. I think you'd like it."

In the relaxed atmosphere of the public house, one of many in the neighbourhood, we're on first-name terms. The *Empire Harmony*, the ship I'd sailed on in the Mediterranean, comes up in the conversation. "Yes," he says. "I noticed you'd sailed on her. I knew Jimmy, your chief on that ship. We sailed together back in the thirties."

It's nice talking to him. His name's Bill. I feel I've made a friend in the office, someone who's been to sea and shares my interests. At least now the company can put a face to my name in their records; I'm not just a number.

It is late afternoon, and the train rocks and clanks across the steel points. A sense of ease comes over me in the passenger compartment. I'm going home, and despite the rain and overcast skies, the day has gone well. The tension drains from my body as I settle back in my seat and glance out of the window. I am at peace with myself.

It's roughly a week into my leave when I pay a visit to Brash Bros. in Rood Lane, just off Fenchurch Street, where I used to work as an office boy. That was in 1940-41, before I started my training as a radio officer. The young women in the office are still as beautiful as when I last saw them, especially Muriel. I used to have a secret crush on her. Eddy, my fellow office boy, is a clerk now.

My old boss, the elderly Miss Irving, still sips from long lines of cups filled with tea. A partner in this firm of tea brokers, Miss Irving is a tea taster in the tea blending department. She savours the sips of tea, then spits them out into a bucket.

A smile spreads across her face when she sees me. She is delighted and greets me warmly. "Gordon, you look wonderful in your uniform," she says. "It's almost three years since you left us. My, you've changed."

That afternoon is burned into my mind because, after I leave the office and turn the corner into Fenchurch Street, I see Karen, Mike's sister. Even though she is dressed in civilian clothes, I still recognize her. She clings to the protective arm of an elderly naval officer, who talks to her as they walk down the street.

It is barely six weeks ago, just before the ship left for Antwerp, that I'd last seen her. She had invited the three of us to a dance at the Women's Auxiliary Army Corps camp where she was stationed. I can still remember how her eyes sparkled when she introduced us to her friends and joked with her brother.

Now, even from across the street, I can see that they are red and swollen from crying. Emotions tug at me, bringing back painful memories. I want to go over and say how sorry I am, but I hesitate. I just can't face her, because I know I will not be able to stop the tears already forming in my eyes.

Sick at heart, I turn away, fighting my emotions. She has enough trouble without me adding to her sorrow. I would only bring back fresh memories of Mike.

"You have that many clothing coupons, Gordon?" My sister Eileen's voice reflects her astonishment as she gazes at the book of coupons given to me by the Ministry of War Transport. The coupons are standard issue to Merchant Navy crews on survivors' leave to replace lost and damaged clothing, uniforms and shoes.

I laugh as I look at her. She is young and pretty, about twenty-two or three. "Can you use some of them, Eileen?" I ask, knowing full well her answer. Her eyes sparkle as I hand her the book.

"Leave me a few for last-minute shopping. I've got most of my things—bridgecoat, shoes, uniforms and odds and ends—so you and Mother might just as well use them if you can."

"Mum," she calls excitedly. "Can we use these coupons? Will the dress shops accept them?"

Mother looks at the coupons carefully, then says, "Yes, I don't see why not; they're not marked for special issue only." A smile creases her worn face as she nods. "Let's go out shopping tomorrow."

Those coupons go fast because clothing is strictly limited and rationed in Britain. There are not many remaining when I finally get the book back.

The forest is damp and swampy after the rains, and the mud squelches underfoot as I pick my way through the quagmire churned up by horses on the bridle path. Dressed for the weather, I draw up the collar of my lightweight raincoat. Low, threatening rain clouds scud across the sky, driven by prevailing westerly Atlantic gales. They are in sympathy with my anger and despair. Yesterday, on a visit to the Radio College in Holloway, I learned from Mr. James, the principal, that two more friends have been killed from our group of five who graduated in 1942.

A bespectacled, greying man in his late forties, the principal sat at his desk in the sparsely furnished office. An enlarged photograph of a younger man in a radio officer's uniform, and a variety of framed photos of ships adorn the walls, leaving no doubt that this was, indeed, a younger Mr. James. A large student record book lay open on the desk in front of him as we talked.

Shifting in his chair, he glanced down at the book as if for reassurance before looking up at me. "Yes," he said in a low voice. "Chris Hamilton was killed in a naval vessel off Sicily in 1943. I think you know he joined the Royal Navy soon after leaving us?"

I nodded as he passed the record book to me. "Yes, we kept in touch for awhile, but you know how it is; we drifted apart. I lost touch in 'forty-three when I went out to the Mediterranean. He never wrote back after awhile. Same with Bob Oakley, I don't know what happened to him."

I stare at the small passport photo of Chris pasted next to the entry. His round cherubic features stare up at me, unwinking, masking his outgoing personality. I felt a pang of pain pass through me. This is not the Chris I knew. The fun-loving teenager is now reduced to an image in a photograph.

"Bob's further on. I'm afraid he died, too, died in Algiers harbour." The principal's voice is soft and hesitant, because he knew these young men and many more who have died.

I cannot speak, but turn the pages until I come to Bob's name. There is a footnote added: *He died in the sea after diving overboard from the burning tanker. The flames from the burning oil floating on the surface swept over him.*

My eyes turn back to the principal as I hand him the book. "There's only Eric and I left," I said, my voice shaking. "Our group of five were very close. John, John Cummings died first, drowned in the Bay of Biscay. I know his parents and heard about it, but then you'll have his record, too, I guess."

The principal's grey eyes never wavered from my face as he said, "Yes, and yours. Torpedoed in the Atlantic in 1942, then sunk in the Scheldt, just a month ago. Our old boys keep us informed. It all goes in here." He taps the record book. "Such a waste of young lives. We train you and then send you out to be slaughtered. It doesn't seem right," he muttered heavily. I left him in his office alone with his thoughts and made my way home.

Now, as I sit on a fallen tree trunk, a confusion of fleeting, flickering images parade through my mind. I smile as I see the five of us in that sleazy little truckers' cafe near the college. Memories surface of cycling on icy roads through rain and snow. I cycled twenty miles a day, ten miles each way, but it was worth it. My smile fades as I remember: most of us are dead. Only Eric and I are left, and so far he has come through without incident.

The crack of a breaking limb brings me back to the present, and for a moment I don't know where I am. My depression lifts momentarily as I recognize this glade. One of my secret dens was here, but now there's just a tangled mass of thorn bushes and brambles. I smile as I remember making that den. It had its own entrance tunneled into the heart of the bushes through a hole so small that only a child could get through. I used to play there alone in the long warm summers, pretending that the forest was populated with imaginary tribes, and that I was a child Robinson Crusoe.

Wandering on, I cut across the wide forest belt towards Connaught Waters, and pause beside a stream. It's peaceful in the forest. Only the sounds of rushing water and the wind in the leafless branches break the quiet. A grey squirrel scurries away, disturbed by the sway of branches in the gusting wind.

I don't know why I crossed the plain to the Royal Forest Hotel when I reached the lake. On impulse I stop in the foyer of the hotel to read the notice of coming events. It was in this hotel ballroom that I used to meet a small group of friends on Saturday evenings. I smile, amused, because I can only remember the young women: their names were Joan, Kathleen and June. I make a note of the ballroom schedules.

I recognize Joan; she's just as I remember. She laughs, tossing back her black, shoulder-length hair, when she sees me. "Gordon," she says, "after all this time! Three years, and look at you. Where's the kid of sixteen?"

"We've all changed, Joan." Pausing, my eyes take in the oval elfin face and large brown eyes; she's not beautiful, but she's an attractive woman with a trim figure, and I like what I see. Then I add thoughtfully, "A little older, perhaps, but you've not changed. It's good to see you. I came up on the off-chance to renew old friendships. I'm glad you came."

The band is playing a soft lilting blues number, and I ask her to dance. She's a good dancer and moves lightly in my arms as we glide into a fox-trot. Joan looks up, her eyes shining. "What are you now?" she asks. "You have two wavy bands on your uniform."

"Second radio officer, though, to most seagoing people, I am Sparks." I pause, then, hesitantly, ask her if she's on her own tonight.

Joan laughs, her eyes twinkling. "I'm with Kathleen and June. Why don't you join us?"

"Yeah," I say, trying to suppress the relief in my voice, "I'd love that." I feel the answering pressure of her hand in mine and know that she, too, is glad to see me.

The crowd of dancers disperses as the band takes a break, and I follow Joan, aware of the way she carries herself and the swing of her hips. She's wearing a white silk dress which contrasts with her black hair. The knee-length full skirt flares out as she walks and the low neckline is fastened with a gold brooch. Joan takes pride in her appearance and doesn't spare the expense.

"Hi, Gordon." Kathleen's smile is broad as we join her and June. "I thought that was you dancing with Joan; the naval uniform's a dead giveaway."

Kathleen's a head taller than Joan and June, and talkative; she's lively company. Framed by a dark curly mop of hair swept back from a high forehead, her face is animated as she speaks. June, overawed by her companions, tends to disappear in the background. Pretty and shy, she's the youngest of the three, and only occasionally ventures an opinion.

The old friendship between us, a friendship dating back to early 1941, has been renewed, but with a difference. It marks the beginning of a more intimate relationship between Joan and me.

It's cold but the rain holds off when Joan and I leave the Royal Forest Hotel ballroom, and catch a double-decker bus which deposits us at the foot of her street in North Chingford. We walk slowly to her gate.

The blacked-out streets are behind me now as I take a shortcut across the ridge and forest at the end of Joan's street. Familiar with the countryside, I take the rough path through the scrub bush skirting the forest edge. I had changed into heavy boots before I left the ballroom and put my light dance shoes in a small backpack.

A strange heady feeling possesses me and, for once, all is right with my world. The memory of Joan's body against mine as we kissed goodnight remains locked in my mind. I've arranged to see her tomorrow. I remember her words when I asked if I could see her again. She laughed lightly as we clung together, "Mmmm, yes, I'll think about it," she teased, then relented. "I'd love to, but remember, I work during the week!"

"Are you still with that law firm in Walthamstow?"

"Yes, I'm a legal secretary now."

"Well, tomorrow's Sunday; how about going out somewhere?" I smiled when she accepted.

I feel an attraction towards Joan; she's shown me a new world and I've been seeing her with increasing regularity over the past few weeks. I think I'm falling in love with the family, especially her mother, Mrs. Green. Small and sparrow-like, Mrs. Green, known affectionately as Mrs. G, made me feel like a part of her family from the first time I met her. She's getting old and her face is tired and worn, but it still reflects her lovable personality. It's a loving home and a closely knit family.

One night after returning from a theatre in London's West End, Joan talked about her mother. It made me realize the strength of that seemingly frail person.

We were alone in the railway carriage on the steam train from Liverpool Street to Chingford, when I mentioned my father's death back in 1942. "Yes," I said, "Mother's changed since Dad's death. She acts strange these days, but she has always worn that old pilot's helmet. She lost all her hair after Douglas was born."

Pausing, I add hesitantly, "I guess if you lose someone close to you it's bound to affect you."

"Yes, I know," Joan replied, "I lost my father when I was very young. Dad owned a coal delivery business long before the war, until drink, gambling and women ruined him. He lost his business and his wife and family." Joan hesitates a moment, then continues, "My parents were divorced and Mum brought us all up. Ernie and Hilda, the two eldest, helped her, but with six of us it wasn't easy."

Now, as I help Mother with the weekly chore of cleaning and emptying the latrines, I can't help but reflect on how different is the world in which Joan lives. Joan's pleasant, semi-detached house with all the modern conveniences is in a good residential area of suburban London. One flush of the toilet and it's all gone down into the sewage system.

"Christ, Mum," I gasp as I empty the buckets, "do you do this yourself when I'm away?"

"Aye, love," she says, "ever since your dad died. After awhile it's nay so bad; you get used to it."

I hold my breath in disgust as I pour the excrement into the deep hole that we've dug in a corner of the allotment, and shake my head in disbelief. *Bloody hell,* I think to myself, *I'd sooner be at sea.* There, we have flush toilets and showers in the ship's head. There's not even a bathroom in the farmhouse, just a portable metal tub which we used on a Saturday night when we were kids. We climbed into it, one after the other in front of the kitchen fire.

As I empty the pails, Mother scrubs and disinfects them. My nose wrinkles; the stench is overpowering, and even the chemicals that I throw down into the pit don't seem to help. Only after I shovel a layer of earth on top of the mess does the smell abate and I can breathe normally again.

I remember seeing my father doing this chore while he was alive. I look at my mother and marvel; she's treating it like any other chore. Grandfather Dunn had a large house in Yorkshire and my mother had not been brought up to this. The villages there at least had the night soil men to cart away the muck in the middle of the night. *I wonder why my father brought her to this godawful life. Surely, he could have chosen a better place.*

Thinking of my father reminds me about Brassy, the old tramp. He used to call me *young Jim.* Brassy wandered about the countryside with his ferrets and sometimes slept in our barn. He was a strange old man who smelt funny. I look at Mother; she's still disinfecting the pails. "Couldn't you get old Brassy to do this, Mum?" I ask.

She laughs, then says, "Occasionally your dad got him to do it. But old Brassy was found dead one morning in the forest about a year back now. Anyway, you couldn't rely on him! He'd take off for weeks then turn up again out of the blue."

She pauses, her thoughts wandering, then continues, "Well, at least we don't have to carry the drinking water from the spring up near the Owl Inn."

"Oh, yeah," I say, "I remember that. It was just before the war when we started getting water from the Golf Club; I must have been about thirteen years old at the time. Didn't some drunk foul up the spring?"

Mother nods and looks at me, "You've got a good memory, love. That's right, Someone from the bloody pub did their business in it. I'd like to have strangled the bastard." She pauses momentarily, and then adds, "That's what prompted us to get permission to have a standpipe and tap installed near the house on the Golf Club's property."

She sighs and continues, "Your brothers help me with that job now. We use your dad's old wooden yoke to carry the water buckets. Remember the three-gallon cans your dad had?" she asks. "We still use them."

I nod, for I know them well. As a child I would sometimes accompany my father. The spring was in a hollow near the Owl Inn, and used to have a padlocked wooden cover. The cover has long gone, and the spring has been abandoned.

Joan has visited the farmhouse, but we don't stay long; it's depressing in winter. She's not used to country life. The fancy shoes she wears in town are useless out here. Rubber boots or stout walking shoes are not Joan's idea of footwear. I smile as she curses the mud. "Good God, Gord," she says, "How can you stick all this mud?" A grin transforms her face and lights up her eyes. "Now I know why you went to sea."

"Yeah," I say with a laugh, "but have you seen the outhouse? Now that's something."

She grimaces. "Yes, your mother showed it to me in case I needed it. Phew, it stinks; I didn't know that they were still used in England. I'd have to be desperate to use that!"

"Oh, yes," I say, "you'll find them in remote rural areas, but we are only twelve or thirteen miles from the centre of London." I'm silent a moment, then continue, "Now perhaps you can understand why I have to get away."

I stare down at the letter in my hand, and read it again. It is from Siemens, my shipping company. My old chief, Jim Hankley, has asked them whether he, John and I can sail together on our next ship. They want to know if I would like to join him as the second radio officer.

My thoughts are mixed. The three of us had got on well together and the offer is tempting, but remembered hurts, such as those experienced on the LST coming back from Ostende, cut deep. Slights magnify in my mind and I no longer trust men, and their insensitivity. The need to be in charge and in control of my own station is very strong. I want to be chief. I need the boost which this position will give to my self-esteem and sense of worth. Diplomatically, I decline and ask for my own ship.

Mike's death, and now two more deaths in our group drives me within myself and I blindly wander the streets of London. Although I take in the occasional variety show, they do not register. But one movie sticks in my mind like a revolving door; it is *Blues in the Night*, starring Ida Lupino. The music evokes in me feelings of longing and loneliness.

It is with great relief that, on March 8, 1945, I again ship out. I have my wish: the company has promoted me to chief radio officer on a small tanker bound for operations under South East Asia Command (SEAC), Trincomalee. With the war in Europe in its closing phase, we head into another war theatre, the South Pacific and the China coast. A new chapter in my life is about to begin.

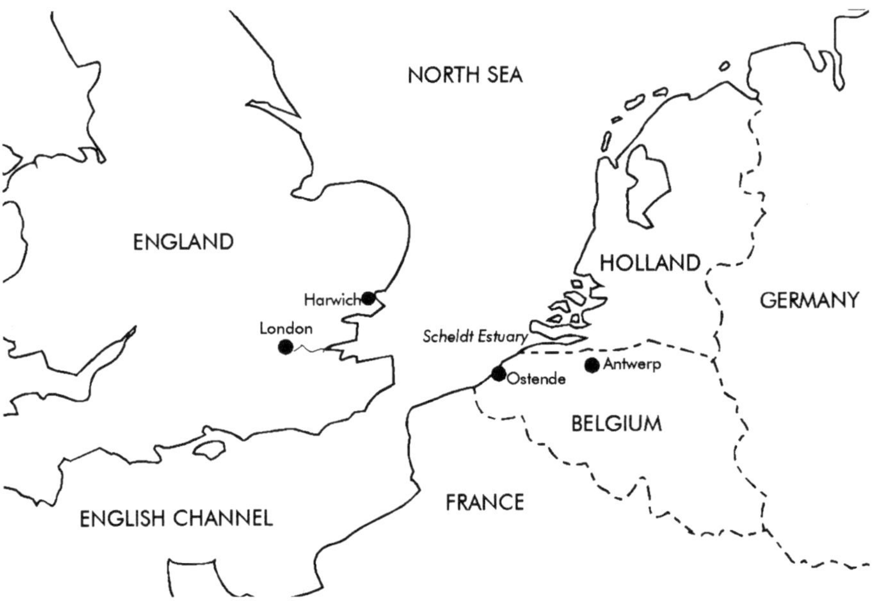

North Sea, showing Scheldt Estuary

AUTHOR'S NOTE

All the events in the book are factual, and are recorded as I remember them. This book describes what happened to the tanker *Scottish Heather*, after she was torpedoed, as well as events affecting the *Empire Harmony* in the North African and Italian campaigns, and the sinking of the *Empire Path* in the Scheldt Estuary. The actual attack on convoy *ONS 154* has been well documented by Henry Revely *(The Convoy that Nearly Died.* London: Wm. Kimber, 1979*)*. Throughout the book, actual names are used for the ships; however, many personal names have been changed.